GI for Disaster Management

GI for Disaster Management

Editors

Orhan Altan
Tullio Joseph Tanzi
Madhu Chandra
Filiz Sunar
Lena Halounová

MDPI • Basel • Beijing • Wuhan • Barcelona • Belgrade • Manchester • Tokyo • Cluj • Tianjin

Editors

Orhan Altan
Istanbul Technical University
Turkey

Tullio Joseph Tanzi
Université Paris-Saclay
France

Madhu Chandra
Professur für
Hochfrequenztechnik and
Electromagnetic Theory
Technical University
of Chemnitz
Germany

Filiz Sunar
Istanbul Teknik Universitesi
Turkey

Lena Halounová
Czech Technical University
in Prague
Czech Republic

Editorial Office
MDPI
St. Alban-Anlage 66
4052 Basel, Switzerland

This is a reprint of articles from the Special Issue published online in the open access journal *ISPRS International Journal of Geo-Information* (ISSN 2220-9964) (available at: https://www.mdpi. com/journal/ijgi/special_issues/GI_disaster_management).

For citation purposes, cite each article independently as indicated on the article page online and as indicated below:

LastName, A.A.; LastName, B.B.; LastName, C.C. Article Title. *Journal Name* **Year**, *Article Number, Page Range.*

ISBN 978-3-03936-824-2 (Hbk)
ISBN 978-3-03936-825-9 (PDF)

Contents

About the Editors

Orhan Altan ALTAN is Em. Prof. at the Istanbul Technical University, Turkey. He is member of several international societies. He served as member of the Executive Board of the ICSU (International Council for Science) during 2011–2018 and has served as Guest Professor in Stuttgart, Berlin, Munich Technical Universities (Germany), ETH-Zurich (Switzerland), and the State Key Laboratory of Information Engineering in Surveying, Mapping and Remote Sensing of Wuhan University (China). He was ISPRS Congress Director in 2004 in Istanbul and has worked as Secretary General during 2004–2008, was President in 2008–2012, and served as 1st Vice-President of ISPRS during 2012–2016. He is an Honorary Member of ISPRS, Honorary Member of Science Academy in Turkey, and Honorary Fellow of The Indian Remote Sensing Society.

Tullio Joseph Tanzi Tanzi is Professor in Institut Mines-Telecom - Telecom ParisTech.Télécom ParisTech - LabSoC, c/o EURECOM. His research interests lie in the areas of network communication, risk management, network architecture, safety, network, disaster management, natural disasters, IT systems, and radio science. He is currently Vice-Chair of the Commission F Wave Propagation and Remote Sensing (planetary atmospheres, surfaces, and subsurfaces) of URSI and chair of the ISPRS Joint Working Group ICWG III/IVa: Disaster Assessment, Monitoring and Management, which is a joint initiative of URSI and ISPRS.

Madhu Chandra is currently the Chairholder and Head of Microwave Engineering and Electromagnetic Theory, Technische Universität Chemnitz, Germany. His research interests include 1. Advanced multi-parameter radars employing diversity of polarization and waveforms in combination with smart and configurable antennas; 2. EM wave propagation and scattering for radar applications; 3. Remote sensing with advanced weather radars and SAR systems; 4. Physics-based signal processing of radar signals for obtaining target information; 5. EM wave propagation for wireless links; 6. Engineering education. His current projects include radar remote sensing, training of radar engineers, and development of radar signal processing algorithms for Doppler and polarization diversity radars.

Filiz Sunar is Lecturer in the Geomatics Department of the Civil Engineering Faculty at Istanbul Technical University. The main positions held include i) Co-director of ITU-CSCRS Satellite Ground Receiving Station, June 2003–2009; ii) ISPRS TC VII (Thematic Processing, Modelling and Analysis of Remotely Sensed Data) president for 2012–2016; iii) Evaluator and Project Monitoring Committee member of the GÖKTURK-2 Project (first national Earth observation satellite) supported by TUBITAK; iv) Evaluator of numerous EU FP7/H2020-EO projects; v) Guest lecturer at Graz University, Padova University, and Politechnico di Milano; vi) Associate Editor of PE&RS Journal (since November 2017).

Lena Halounová Since graduation from the CTU in Prague (Faculty of Civil Engineering), Prof. Dr. Lena Halounová has remained within its walls during her Ph.D. at the Department of Hydrotechnics before moving to the Department of Geomatics. In her research, Lena pays special attention to issues of using optical and SAR remote sensing data and GIS applications for solving problems of water engineering, erosion, reclamations, landslides, land subsidence, detection of vegetation in urban areas, and change detection in urban areas. Lena was elected the Director of the 2016 ISPRS Congress, where she was elected ISPRS Secretary General for the 2016–2020 period.

International Journal of
Geo-Information

Editorial

Spatially Supported Disaster Management: Introduction to the Special Issue "GI for Disaster Management"

Orhan Altan [1,*], Tullio Joseph Tanzi [2], Madhu Chandra [3], Filiz Sunar [1] and Lena Halounová [4]

[1] Department of Geomatics, Istanbul Technical University, 34469 Istanbul, Turkey; fsunar@gmail.com
[2] LTCI, Telecom Paris, Institut Polytechnique de Paris, 75013 Paris, France; tullio.tanzi@telecom-paris.fr
[3] Microwave Engineering and Electromagnetic Theory, Chemnitz, Technical University of Chemnitz, 09126 Chemnitz, Germany; madhu.chandra@etit.tu-chemnitz.de
[4] Civil Engineering, Czech Technical University in Prague, Thákurova 7, 166 29 Prague, Czech Republic; halounov@gmail.com
* Correspondence: oaltan@itu.edu.tr

Received: 30 April 2020; Accepted: 4 May 2020; Published: 8 May 2020

This special issue explores most of the scientific issues related to spatially supported disaster management and its integration with geographical information system technologies in different disaster examples and scales. The need for a detailed use of geoinformation in disaster management is a fact [1,2]. Dealing with disasters over space and time represents a long-lasting theme, now approached by means of innovative techniques and modelling approaches (Gi4DM Conference Series, http://www.gi4dm.net/). Several priorities for actions to reduce existing disaster risks and prevent new ones include understanding disaster risk, strengthening disaster risk governance for the management of disaster risk, investing in disaster reduction for resilience, and enhancing disaster preparedness for effective responses [3,4] are outlined.

This special issue explores some of them, with challenging ideas facing different components of spatial patterns related to ecological processes and the published articles are selected and extended versions from the Gi4DM Conference in 2019 in Prague, Czech Republic. Articles [5–7] deal with sea-level rise. In the last decades, forest fires became a major disaster phenomenon and Luis Paduva [8] developed some innovative techniques such as UAV-borne observations and compared them with sentinel data. During crisis response, it is critical to share and understand complex spatial, thematic, and temporal information in a timely, visual and compelling way. Cartography plays an important role in delivering reliable, understandable, appealing and user-friendly visual information through maps. In order to provide seamless communication between heterogeneous audiences at the time of a disaster, Kuveždić Divjak [9] deals with a unique environment for cartographic symbolization. A very interesting study dealing with suitable site selection and planning of urban areas affected by multiple hazards, and their integration into hazard susceptibility maps, can be seen in Yanar [10]. Norman Kerle and his colleagues deal with one of the oldest remote sensing challenges, i.e., structural disaster damage detection and characterization [11].

ISPRS are working closely with many GeoUnions, such as URSI (Union Radio Science International), an organization bringing together experts from remote sensing and geoinformation technology and related sciences and technology. Papers from various events—congresses, Gi4DM conferences, several symposia workshop—are available at https://www.isprs.org/publications/Default.aspx.

This special issue gathers papers from the best researchers in the field who properly faced different aspects of spatial ecology, including problems and uncertainties which in most cases remain unaddressed in disaster management, at large.

We hope that the readership of the ISPRS International Journal of Geo-Information will enjoy the scientific effort put into properly facing and finding solutions to very different issues of disaster management.

References

1. Altan, O.; ToZ, G.; Kulur, S.; Seker, D.; Volz, S.; Fritsch, D.; Sester, M. Photogrammetry and geographic information systems for quick assessment, documentation and analysis of earthquakes. *ISPRS J. Photogramm. Remote Sens.* **2001**, *55*, 359–372. [CrossRef]
2. Altan, O. Use of Photogrammetry, Remote Sensing and Spatial Information Technologies in Disaster Management, especially Earthquakes. In *Geo-Information for Disaster Management*; Van Oosterom, P., Zlatanova, S., Fendel, E.M., Eds.; Springer: Berlin/Heidelberg, Germany, 2005. [CrossRef]
3. Ismail-Zadeh, A.; Cutter, S. (Eds.) *Disaster Risks Research and Assessment to Promote Risk Reduction and Management*; International Science Council: Paris, France, 2015. Available online: http://www.iugg.org/policy/Report_RiskReduction_WCDRR_2015.pdf (accessed on 15 April 2020).
4. Cutter, S.; Ismail-Zadeh, A.; Alcántara-Ayala, I.; Altan, O.; Baker, D.N.; Briceño, S.; Gupta, H.; Hollaway, A.; Johnston, D.; McBean, G.A.; et al. Pool knowledge to stem losses from disasters. *Nature* **2015**, *522*, 277–279. [CrossRef] [PubMed]
5. Baučić, M. Household Level Vulnerability Analysis—Index and Fuzzy Based Methods. *ISPRS Int. J. Geo Inf.* **2020**, *9*, 263. [CrossRef]
6. Avşar, N.B.; Kutoğlu, Ş.H. Recent Sea Level Change in the Black Sea from Satellite Altimetry and Tide Gauge Observations. *ISPRS Int. J. Geo Inf.* **2020**, *9*, 185. [CrossRef]
7. Nirwansyah, A.W.; Braun, B. Mapping Impact of Tidal Flooding on Solar Salt Farming in Northern Java using a Hydrodynamic Model. *ISPRS Int. J. Geo Inf.* **2019**, *8*, 451. [CrossRef]
8. Pádua, L.; Guimarães, N.; Adão, T.; Sousa, A.; Peres, E.; Sousa, J.J. Effectiveness of Sentinel-2 in Multi-Temporal Post-Fire Monitoring When Compared with UAV Imagery. *ISPRS Int. J. Geo Inf.* **2020**, *9*, 225. [CrossRef]
9. Kuveždić Divjak, A.; Đapo, A.; Pribičević, B. Cartographic Symbology for Crisis Mapping: A Comparative Study. *ISPRS Int. J. Geo Inf.* **2020**, *9*, 142. [CrossRef]
10. Yanar, T.; Kocaman, S.; Gokceoglu, C. Use of Mamdani Fuzzy Algorithm for Multi-Hazard Susceptibility Assessment in a Developing Urban Settlement (Mamak, Ankara, Turkey). *ISPRS Int. J. Geo Inf.* **2020**, *9*, 114. [CrossRef]
11. Kerle, N.; Nex, F.; Gerke, M.; Duarte, D.; Vetrivel, A. UAV-Based Structural Damage Mapping: A Review. *ISPRS Int. J. Geo Inf.* **2020**, *9*, 14. [CrossRef]

International Journal of
Geo-Information

Article

Household Level Vulnerability Analysis—Index and Fuzzy Based Methods

Martina Baučić

Faculty of Civil Engineering, Architecture and Geodesy, University of Split, Matice hrvatske 15, 21000 Split, Croatia; martina.baucic@gradst.hr

Received: 31 January 2020; Accepted: 17 April 2020; Published: 19 April 2020

Abstract: Coastal vulnerability assessment due to climate change impacts, particularly for sea level rise, has become an essential part of coastal management all over the world. For the planning and implementation of adaptation measures at the household level, large-scale analysis is necessary. The main aim of this research is to investigate and propose a simple and viable assessment method that includes three key geospatial parameters: elevation, distance to coastline, and building footprint area. Two methods are proposed—one based on the Index method and another on fuzzy logic. While the former method standardizes the quantitative parameters to unit-less vulnerability sub-indices using functions (avoiding crisp classification) and summarizes them, the latter method turns quantitative parameters into linguistic variables and further implements fuzzy logic. For comparison purposes, a third method is considered: the existing Index method using crisp values for vulnerability sub-indices. All three methods were implemented, and the results show significant differences in their vulnerability assessments. A discussion on the advantages and disadvantages led to the following conclusion: although the fuzzy logic method satisfies almost all the requirements, a less complex method based on functions can be applied and still yields significant improvement.

Keywords: climate change; fuzzy logic; GIS, household; Index method; sea level rise; vulnerability

---

1. Introduction

At present, the assessment of coastal vulnerability due to climate change impacts is an essential input for coastal management processes all over the world [1]. The impact of sea level rise is the main climate impact taken into account when managing coastal areas, from determining future use to planning adaptation measures [1–3]. The term "vulnerability" is a function of hazard characteristics, the sensitivity of the assets exposed, and adaptive capacity, which all vary by time and depend on contexts such as socio–economic factors [2]. Hazard characteristics define the exposure of the system to phenomena, sensitivity describes how the system is affected, and adaptive capacity defines the system's ability to maintain its functions. For example, in the case of a flood, the hazard characteristics are water depth and velocity; sensitivity is represented by the number of people and assets flooded; and adaptive capacity is the capacity of emergency infrastructure and flood defence structures. In this paper, "vulnerability" refers to physical vulnerability, as described above, from the perspectives of disaster management, climate change, and other related aspects [4]. Sociology and economics refer to "social vulnerability", which focuses on identifying the most vulnerable groups of people and examines social factors and economic assets, such as poverty, access to food and housing, and human and social capital [4,5]. A vulnerability assessment could follow a quantitative approach based on related indicators and indices, or a qualitative approach based on the perspectives of stakeholders [6]. In this paper, a quantitative approach based on indices is used.

Vulnerability requires assessment methods that apply to different scales: spatial (large, medium, and small), temporal (short, mid, and long term), and management (local, regional, and national) [1].

While assessments used at the national or regional level identify the most vulnerable areas and help in prioritizing measures (scales in the range of 1:100.000 to 1:25.000), large-scale assessments are necessary for the planning and implementation of adaptation measures at the household level [3,7]. An overview of the methods for assessing coastal vulnerability is given in [1]. These methods are classified into four categories: Index-based methods, Indicator-based approaches, GIS-based decision support systems, and methods based on dynamic computer models.

Index-based and Indicator-based methods differ in the methodological approaches they use for the definition of indices/indicators, but are similar in their quantification and combination of indices/indicators in a single parameter describing vulnerability.

Index-based methods calculate the vulnerability index (unit-less value) by summarizing sub-indices (values of selected parameters) [8,9]. This method is widely recognised and has several variants, such as the coastal vulnerability index for sea level rise [10], the composite vulnerability index [11], and the multi-scale coastal vulnerability index [12]. Index-based methods start with a selection of key parameters that represent the processes or assets of importance for coastal vulnerability. The physical parameters include geomorphology, shoreline change rates, coastal slope, rate of sea level rise, wave heights, tidal range, and proximity to coast. The human influence parameters are river regulation, engineered frontage, land use, coastal protection structures, etc. The socio–economic parameters are the affected population, the affected cultural heritage, the affected infrastructure, etc. The second step quantifies the contribution of the selected key parameters to coastal vulnerability using indices from 1 to 5, where 1 indicates a low contribution to coastal vulnerability, and 5 indicates a high contribution to coastal vulnerability. Experts have developed classification schemas, where the key parameter values are classified by the values of the sub-indices. For example, a sea level rise with a rate of less the 1 mm/year is classified as value 1 for the sub-index, that with a rate of 1–2 mm/year is given value 2, that with a rate of 2–5 mm/year is value 3, that with a rate of 5–7 mm/year is value 4, and that with a rate of 7 mm/year and over is value 5 [1]. The final step integrates the sub-indices into a single index via the use of selected formulas, such as the product mean or average sum. Additional refinements could be done by using weights for the sub-indices. Index-based methods are used for various scales, from the small scales used at the national level to the large-scales used at the local level.

Indicator-based approaches use indicators representing coastal vulnerability factors, such as sea level rises; extreme weather conditions; coastal erosion and accretion; and the natural, human, and economic assets at risk. Indicators are defined and quantified primarily to measure the progress towards sustainable development, and to guide decision makers in managing coastal areas. Thus, these indicators differ from the indices used by Index-based methods that focus on vulnerability. Similar to Index-based methods, the indicators can be further classified according to their contribution to coastal vulnerability and integrated into a single indicator of vulnerability. For example, the Deduce Interreg project developed a set of 27 core indicators for sustainable coastal zone development [13]. Three indicators addressed climate change vulnerability—(i) sea level rise and extreme weather conditions (quantified by the number of stormy days, the sea level rise, and the length of the protected coastline); (ii) coastal erosion and accretion (quantified by the length of the dynamic coastline, the area and volume of sand nourishment, and the number of people living in the coastal flooding areas); and (iii) the natural, human, and economic assets at risk (quantified by the areas of the protected sites and by the economic values of the assets in the coastal flooding areas). Indicator-based approaches are used for national, regional, and local level assessments.

More complex methods, such as GIS-based decision support systems or dynamic computer models, fit a particular study area and use comprehensive data sets (3D models) and engineering applications. An example of a GIS-based decision support system is DESYCO [14]. This system evaluates various climate change impacts and implements a Regional Risk Assessment methodology based on Multi-Criteria Decision Analysis. Methods based on dynamic computer models either focus on a particular coastal process (e.g., the RACE approach, focusing on coastal erosion [15]) or provide integrated assessments of the regional and national levels (such as DIVA [16]). Coastal engineering applications, such as the Delft3D modelling suite [17], use 3D models that can be applied to coastal vulnerability assessment.

To facilitate the selection of the most appropriate method, Ramieri et al. [1] summarized the advantages and disadvantages of the above described methods as follows. Index and Indicator-based methods are simple to implement and appropriate for vulnerability scoping, while GIS-based decision support systems and dynamic computer models provide detailed quantitative assessments and the identification of adaptation measures. The disadvantages of the GIS-based decision support systems and dynamic computer models are their high requirements for data and expert knowledge. Furthermore, complex methods are not easily understood by the public, thus making it difficult to raise awareness and motivate owners to start such adaptation measures [1]. Even for Index methods, the final index is not transparent, because it encapsulates various assumptions, generalisations, calculations, etc. Miller et al. [18] elaborated the challenges that always remain as the following: the selection of representative variables for the study area, the definition of weights for the indicators, the availability of data, and the validation of the results.

The current state of the research into vulnerability assessment can be grouped according to the following main objectives:

- Answering the needs of particular applications: the modification of methods and models to fit particular geographic areas and study needs (e.g., by introducing new parameters and models for their evaluation) [1,10–13,18–21];
- Improving existing methods: the development of more complex and sophisticated methods using Fuzzy Logic, Analytical Hierarchy Processes, and similar models [14–17,22–34];
- Supporting coastal management: including more socio–economic issues, decision making processes, or public awareness [33,35,36].

Vulnerability assessments at the regional level are not detailed enough to be included in coastal area land use planning or in planning adaptation measures, particularly in urban areas [3,7,19]. In order to enable local authorities to include climate change impacts in coastal area management activities, there is the need for a method at a large-scale level to assess each building, hereafter called the household level method. This assessment should include economic aspects, such as building damage, social aspects, such as population vulnerability, and adaption measures at the household level [4,5,37–41].

The author's previous work on several coastal and flood vulnerability assessment projects has led to this research and is summarized as follows. For the regional level analysis, the Index method was used and applied to coastline segments at a scale of 1:25.000 [42] or to areas at a scale of 1:100.000, with raster tessellation of a 100 x 100 m pixel size [43]. The latest work featured a large-scale analysis for the coastal area of the City of Kaštela [44,45]. The most valuable assets in this study, such as residential and historical buildings, were located along 23 km of the coastline and were already prone to coastal flooding. A vulnerability assessment was undertaken to support the development of priorities and measures for the coastal action plan of the City of Kaštela. There was a need for an initial vulnerability assessment for each building in the coastal zone, and thus an index-method was selected. Indicator-based methods have a much wider focus, and elaborate upon dozens of indicators; thus, they could be used to measure the progress towards sustainable development. GIS-based decision support systems and dynamic computer models have high requirements for data and expert knowledge. Such models could be used in future work to provide detailed engineering solutions for the selected locations. The vulnerability index was calculated for each building by summarizing the sub-indices. The four sub-indices were calculated based on the parameters describing exposure to hazards (location of buildings in hazard zone 1, 2, or 3) and sensitivity variables (building usage, building temporal usage, and the building's construction status). By using an Index-based method for the large-scale assessment, the following questions emerged:

1. What spatial units or tessellation types should be used for large-scale vulnerability assessment?
2. How can we deal with the uncertainties in vulnerability assessment?

3.   How can we use Index and Indicator-based methods with crisp classifications when key variables represent continuous phenomena?
4.   How could vulnerability be more easily accepted by local planners and society?

Thus, the relevant work describing methods for vulnerability assessment at a large-scale level was analysed according to the above questions [7,25,27,32,36–41,46,47]. A short discussion follows.

Hazard characteristics, sensitivity of the assets exposed, and the adaptive capacity are all represented by various spatial features. These factors are combined into the final spatial features with homogenous vulnerability values. An approach used in medium- and small-scale assessments is the following. Spatial units that represent vulnerability assessments are often administrative units, such as provinces, municipalities, settlements, city blocks, or statistical units. The aggregate values of key parameters are calculated and assigned to selected spatial units, such as the number of inhabitants in each city block. For continuous phenomena, such as water depth, these aggregation values (e.g., the average water depth for each city block) introduce certain uncertainties into the assessment, because the key parameter values are not homogeneous over all the area covered by the spatial unit. Regular spatial tessellation is used as well, where the cells of adequate sizes are assigned key parameter values, and the final vulnerability assessment is calculated, but the same cause of uncertainty remains.

For large-scale assessments, the final vulnerability assessment can be assigned to the regular spatial tessellation units of a small size, e.g., 1×1 m. Large-scale assessments are fine enough to distinguish particular assets and thus another approach could also be used: assigning vulnerability to each object representing affected assets, such as infrastructure objects (roads, utilities) or buildings. Several studies focused on the identification of affected buildings and the calculation of economic losses because buildings are key assets for people [7,37–39]. Adaption measures, the calculation of damage to physical objects, and socio–economic vulnerability parameters, such as household income and unemployment, are all spatially assigned to the buildings.

Regarding the uncertainties in vulnerability assessment, the data can be vague, as can the models and problem definitions, but there is also subjectivity in making decisions [36,46,47]. For example, the spatial representation of vague phenomena, such as floods, using polygons with well-defined boundaries introduces errors in the assessments [25,37]. Experts, together with decision makers and other involved participants, must also quantify the contributions of key parameters and thus introduce a certain level of subjectivity [36]. For the thematic aspects, Index- and Indicator-based methods for vulnerability assessment include the crisp classification of parameters, although there are uncertainties. Finally, vulnerability, when represented by polygons with assigned vulnerability indices, encapsulates the uncertainty of the spatial extent and vulnerability indices. Jadidi et al. [25] developed a diagram of spatial uncertainty and the methods to handle it. The nature of uncertainty lies in its epistemic descriptions given by measured or sampled data, or in its ontological descriptions given by feature definitions that can be well or ill defined. Each spatial feature has its own position, geometry, and description that can be uncertain. One of the methods for modeling this uncertainty is fuzzy set theory, which can model continuous and heterogeneous phenomena [25]. Fuzzy set theory offers a model for "fuzziness", and was introduced in 1965 as an extension of Boolean set theory [48]. Regarding the spatial aspects, there is a need to include vagueness in the definition of boundaries, which is not feasible when using standard vector representations, such as polylines. Therefore, the concept of Fuzzy Spatial Data Types with accompanying Fuzzy Spatial Set Operations and Fuzzy Topological Predicates is introduced [27].

Key parameters describing the vulnerability aspects could be continuous or discrete spatial phenomena. Index- and Indicator-based methods use crisp classifications to quantify the contributions of key parameters. Physical parameters are often continuous phenomena, such as elevation, slope, wave heights, or proximity to coast. Thus, the crisp classification of the vulnerability indices from 1 to 5 introduces crisp boundaries, and there is no transition from one vulnerability level to another. In the case of a flood, exposure to the flood could be represented by polygons and classified with a vulnerability level from 1 to 5 based on its elevation from 1 to 5 m. Thus, two buildings with elevations of 0.1 and 0.9 are ranked with 1, and the building with an elevation of 1.1 is ranked with 2, which is not a realistic representation, since exposure to flood changes gradually. One of the proposed approaches is to use fuzzy rule-based classification. The statistical study of whether there is a significant statistical difference in performance between crisp and fuzzy rule-based classification has confirmed that these two classification methods offer the same statistical meaning [32]. In this work, conversely, testing the methods using (among other steps) crisp and fuzzy classifications resulted in significant differences.

In order to support coastal management, particularly the implementation of adaptive measures at the household level, the proposed method for large-scale assessment should be easily accepted by local planners and society. From the author's experience, such a method should emphasize the following:

- The use of existing data (locally/nationally available or open data sets);
- The use of available tools (common tools, such as a spreadsheet programs and open source GIS tools);
- Simplicity, ease of understanding, and the ability to be implemented by coastal planners and managers of various levels of expertise; and
- Effectively communicate vulnerability to coastal management stakeholders (e.g. local authorities, utility companies, public).

To conclude, there is a need for a method that can provide answers for the above questions. The relevant work proposes using buildings as spatial units for vulnerability assessment and fuzzy set theory for resolving the uncertainty and pitfalls of crisp classification. This research investigates the adaption of Index-based methods by using continuous ranking and fuzzy logic. This research was narrowed to the main climate change impact of sea level rise [1], and to the three key geospatial parameters of elevation, distance to the sea, and building footprint area, which are universal for all geographic areas and essential for vulnerability assessments [3–7,13,18,37,40,41,49]. Thus, two new methods are proposed:

1. The Index method: continuous ranking by functions; the modified Index method uses functions and assigns continuous values to sub-indices;
2. The fuzzy logic method: ranking by membership functions; the modified Index method uses fuzzy logic membership functions, rules, and calculates conclusions.

A third method is implemented for the purpose of the analysis:

1. The Index method: crisp ranking by scores, as described in the literature, using crisp values for the sub-indices.

The newly proposed methods overcome the pitfalls of crisp classification. While the first method standardizes the quantitative parameters to unit-less vulnerability sub-indices via functions and summarizes them, the second method transforms quantitative parameters into linguistic variables, and further implements fuzzy logic in a way that is easily repeated by nonexperts, while still improving the common understanding of the assessment. Both methods use building footprints as crisp features, defined with crisp borders (polygons) as entities to which the vulnerability indices are assigned. To overcome the continuous nature of geospatial parameters, elevation, and distance to the sea, these parameters are not classified in crisp zones, but their values are assigned to the building footprints and then classified by functions or fuzzy logic membership functions. Thus, these methods avoid the complex implementation of Fuzzy Spatial Data Types or similar concepts. The proposed fuzzy logic method includes a definition of the linguistic variables for evaluation of the input parameters (key geospatial parameters) and for the final result. Therefore, this method accommodates technical concepts and their definitions using semantics that are comprehensible by coastal managers and the public.

The results of all three methods are compared, and conclusions are drawn. The final aim of this research is to propose a simple method for coastal vulnerability assessment at the household level that could be widely and easily used and, as a final aim, to provide support to coastal management in the context of climate change impacts.

2. Materials and Methods

Two newly proposed methods and an existing method are implemented in this study. The starting point is a selection of key geospatial parameters. Based on this selection, geospatial data sets are created, and the parameters are calculated for each building. For each implemented method, there are three common steps. The first step performs a ranking, the second step performs calculations of the sub-indices or, for the fuzzy logic method, defines the rules and offers a final conclusion. The third step calculates a single vulnerability index for each building. Finally, the results are compared. Figure 1 depicts these steps and the following paragraphs briefly describe them.

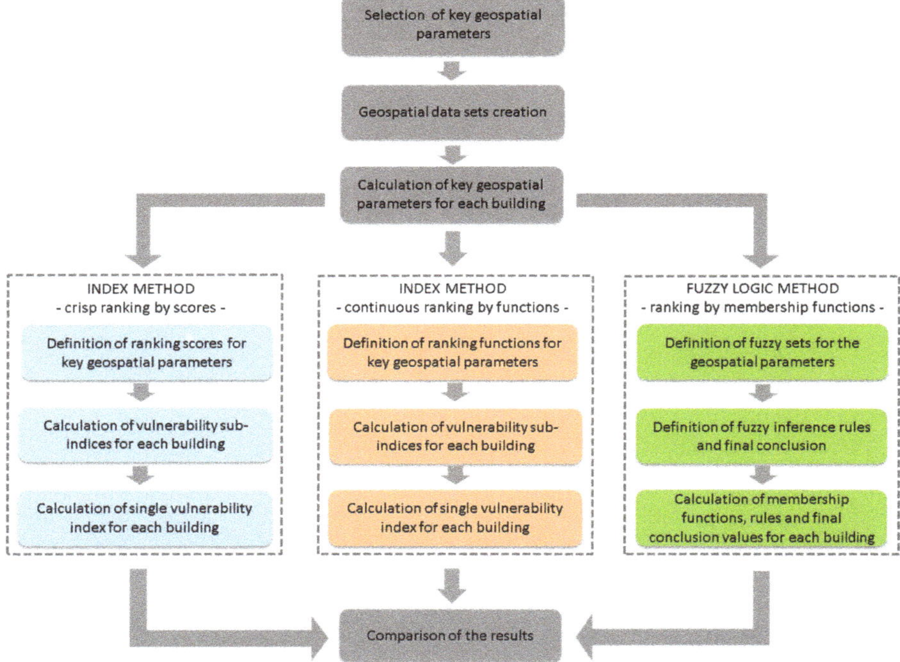

Figure 1. Research steps.

2.1. Selection of the Key Geospatial Parameters

A literature study was carried out to identify the geospatial variables used for coastal vulnerability assessment, and to further select the variables of key importance for sea level rise and household level analyses. For a household level assessment, Miller et al. [18] attempted to reduce the number of indicators to only the most relevant one. Their conclusion was that physical exposure is more important than social characteristics. In the case of coastal flooding, elevation and distance to the coastline describe the exposure to hazards, while the building's footprint areas describe buildings, which are the key assets for people [7,37–39]. Thus, the selected parameters are the following:

- The building's footprint area;
- Elevation above sea level;
- Distance to coastline.

Buildings were proposed as spatial units for vulnerability assessment at the household level by several relevant works [3,6,7,13,18,40]. A building's footprint area is used to calculate socio–economic impacts, such as the damage cost and insurance coverage, and to plan adaptation measures in case of flooding. Additionally, socio–economic vulnerability parameters, such as household income and unemployment, are all spatially assigned to buildings, and thus their assessment could be easily extended by any of these parameters.

Two basic parameters describing exposure to coastal floods are elevation and distance to the coastline. Elevation describes the hazards according to the depth of coastal flooding, and thus describes physical characteristics. Distance to the coastline represent a psychical factor, but it also has social importance, because people have the perception of hazards when living in risk-prone areas, which influences their behaviour [3,6,7,13,18,40]. The selected geospatial parameters are not mutually related.

Slope as a geospatial parameter is used in small and medium-scale analyses. However, in large-scale analyses where each building is assessed, slope does not contribute to building vulnerability assessments and is not selected.

2.2. Study Area and Data

The study area covers 110 ha of the coastal area up to 3 m above the mean sea level. It is an urbanized area of the City of Kaštela that stretches for 23 km along the Kaštela Bay, and is situated on the eastern coast of the Adriatic Sea. Valuable historical settlements and sea promenades are situated close to the sea, and the whole study area includes 1657 buildings. In order to evaluate each building, large-scale data are used for the digital elevation model (DEM), the coastline, and building footprints. The aim was to use the data, either globally or nationally, that are available to local authorities.

For the household level assessment, free global digital elevation models do not satisfy our needs because their spatial resolution is too coarse for urban areas [45]. Some European Union (EU) countries have published open DEM data with spatial resolutions of 10 m, 1 m, and even 0.5 m [50], while other EU countries have provided the same resolutions to their local authorities under certain agreements and financing schemas. For the study area, the national DEM data are used. The vertical accuracy of the triangulated irregular network model (TIN) derived from these data is estimated to be ± 0.35 m for more than 85% of the data in urban areas [51]. The TIN model is converted to a raster model with a spatial resolution of 1 m (hereafter, DEM Kaštela).

Coastline and building footprint data from the national map at a scale of 1:5000 are available in digital format and used by national local authorities for spatial planning purposes; thus, they are also used in the study (hereafter, Buildings and Coastline). An alternative data-set could be Open Street Data (OSD), as the study in [52] concluded that the building data from OSD can be considered a valid and accurate data source corresponding to a 1:5000 scale.

2.3. Calculation of the Geospatial Parameters of the Buildings

The open source software QGIS [53] was used to calculate the key geospatial parameters for the buildings. A brief description of the calculations and used functions follows.

The building footprint areas were calculated from the polygons on an individual basis. Figure 2a shows the results. Elevation above the sea level was calculated as the mean value of the elevations covered by the building's footprint. Figure 2b shows the results. Distance to coastline was calculated as the shortest distance from the building's footprint to the coastline. Firstly, the polylines representing the buildings and coastline were converted to nodes, and for each building's node, the distance to the coastline's node was calculated. If necessary, the nodes representing the coastline could be densified. To obtain the shortest distance, the calculated distances from the nodes belonging to one building were grouped, and the minimal value was selected. Figure 2c shows the results of this process. Each step resulted in a new value being stored in the attribute tables of the Buildings.

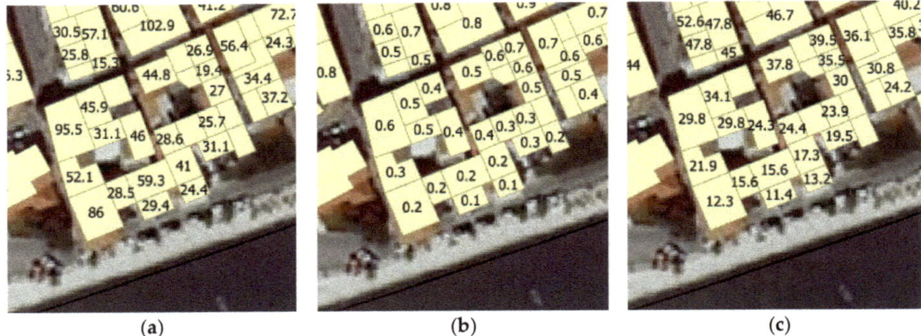

(a)　　　　　　　　　　(b)　　　　　　　　　　(c)

Figure 2. Maps showing the key geospatial parameters for each building in the study area: (**a**) the building's footprint area (m^2); (**b**) the elevation above sea level (m); (**c**) the distance to coastline (m).

2.4. Methods

Three methods are implemented:

- The Index method: crisp ranking by scores;
- The Index method: continuous ranking by functions;
- The Fuzzy logic method: ranking by membership functions.

To implement these methods, particularly to define the rankings, the contributions of the key geospatial parameters to vulnerability should be defined based on an expert evaluation of the study area.

For the City of Kaštela, the study presented in [44,45] defined the hazard zones for coastal flooding as a zone up to 1 m above sea level (already under flooding during storm surges): zone 2 is up to 2 m above sea level, and zone 3 is up to 3 m above sea level. For distance from the coastline, the contribution is defined as high for distances up to 35 m (which have daily and close visual contact with the sea), medium for distances from 35 to 75 m, and low for distances longer than 75 m. For the building's footprint area, the contribution is defined based on the costs of building repair and insurance coverage [49]. The insurance coverage offered on the local market completely covers the building basement repair costs for approximately a 15 m^2 footprint area, and the cost to repair an area of 45 m^2 corresponds to the average annual wage per capita. Thus, a footprint area less than 15 m^2 has a low contribution to vulnerability, that from 15–45 m^2 has a medium contribution, and that larger than 45 m^2 contributes highly to building vulnerability.

A further elaboration of the above defined contributions to vulnerability is not the focus of this research. Moreover, the definitions of the contributions are dependent on the specifics of the study area, and the analytical requirements cannot be generally defined. All three methods were implemented via the open source software QGIS and a spreadsheet calculator.

2.4.1. Index Method—Crisp Ranking by Scores

Using the contributions of the key geospatial parameters given by an expert's evaluation, the vulnerability sub-indices are defined by the use of crisp values for ranking: 5 for high, 3 for medium, and 1 for low contributions (Table 1). The rankings are visualised in Figure 3. The attribute table of the Buildings was exported from QGIS [53], and further calculations were performed in a spreadsheet calculator using the expressions given in Table 1.

Table 1. Crisp ranking by scores: the vulnerability sub-indices V_{area}, $V_{elevation}$, and $V_{distance}$.

Geospatial Parameter	Vulnerability Sub-Index 5 (High)	Vulnerability Sub-Index 3 (Medium)	Vulnerability Sub-Index 1 (Low)	Spreadsheet Expression for Calculation
building's footprint area	>45 m^2	15–45 m^2	<15 m^2	=IF("area"<=15;1;IF("area"<45;3;5))
elevation above sea level	<1m	1–2 m	>2 m	=IF("elevation"<=1;5;IF("elevation"<2;3;1))
distance to the coastline	<30 m	30–75 m	>75 m	=IF("distance"<=30;5;IF("distance"<75;3;1))

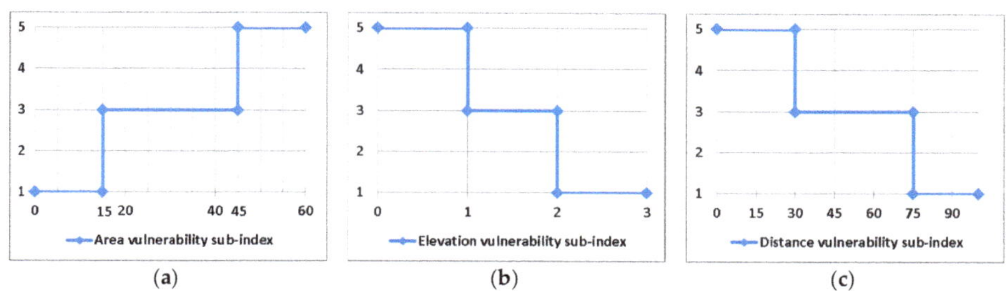

Figure 3. Graphs showing the crisp rankings of the key geospatial parameters—vulnerability sub-indices: (a) for the building's footprint area; (b) for elevation above the sea level; and (c) for distance to coastline.

The definition of a single vulnerability for each building implies that the sub-indices should be integrated into one index called the single or final index. Various relevant approaches have been described and commented upon in the literature, and an overview is given in [1]. As this research was narrowed down to only the key geospatial parameters, and the intention was not to additionally quantify the contributions of vulnerability parameters, the simplest equation was used, featuring the average of the sub-indices (Equation (1)).

$$V = \frac{V_{area} + V_{elevation} + V_{distance}}{3} \tag{1}$$

The calculated single indices are in the range of 1 to 5 and are further rounded to the nearest integers, 1, 2, 3, 4, and 5, representing the building vulnerability index and defining vulnerability using linguistic expressions as follows:

- 1—low vulnerability;
- 2—medium low vulnerability;
- 3—medium vulnerability;
- 4—medium high vulnerability;
- 5—high vulnerability.

2.4.2. Index Method—Continuous Ranking by Functions

Here, the previously implemented Index method was modified such that instead of crisp rankings, continuous rankings were used. Using the contributions of key geospatial parameters to the building vulnerability, the vulnerability sub-indices were defined by using functions and assigning then the following values: 5 for a high contribution and 1 for a low contribution along with continuous values from 1,1 to 4,9 for a medium contribution (Table 2). The ranking functions are visualised in Figure 4. The calculations were done in a spreadsheet calculator using the expressions given in Table 2.

Table 2. Continuous rankings: the vulnerability sub-indices V_{area}, $V_{elevation}$, and $V_{distance}$.

Geospatial Parameter	Vulnerability Sub-Index 5 (High)	Vulnerability Sub-Index 4,9–1,1 (Medium)	Vulnerability Sub-Index 1 (Low)	Spreadsheet Expression for Calculation
building's footprint area	>45 m²	15–45 m²	<15 m²	=IF("area"<=15;1;IF ("area"<45;(4/30*("area"-15)+1);5))
elevation above the sea level	<1m	1–2 m	>2 m	=IF("elevation"<=1;5;IF (("elevation"<2;(-4*("elevation"+9);1))
distance to the coastline	<30 m	30–75 m	>75 m	=IF("distance"<=30;5;IF ("distance"<75;(-4/45*("distance"-30)+5);1))

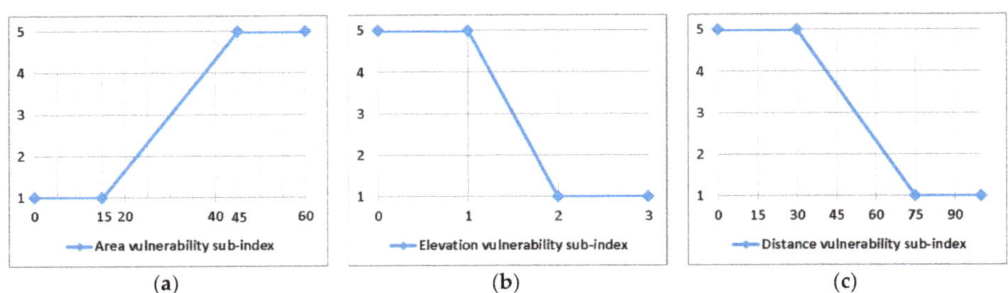

Figure 4. Graphs showing the continuous rankings of the key geospatial parameters—vulnerability sub-indices: (**a**) for the building's footprint area; (**b**) for the elevation above sea level; (**c**) for the distance to coastline.

For a single vulnerability index of each building, the same definitions and calculations used for the Index method were employed, with the crisp rankings given in the previous paragraph (Section 2.4.1).

2.4.3. Fuzzy Logic Method—Ranking by Membership Functions

The Index method was modified such that instead of crisp rankings, fuzzy logic membership functions assigned a membership value to a fuzzy set. In fuzzy set theory, an element's membership to a set is described by its membership function. The membership function values are between 0 and 1, indicating the degree of membership [48] and are described in a linguistic form such as "near" and "far". Moreover, fuzzy logic offers the new concept of integrating the sub-indices into a single index via logical reasoning with a generalized modus ponens (rules of inference), as shown in Equation (2). In this research, a simplified fuzzy logic method is used. Here, the conclusion of a rule is not a fuzzy set but a number. The single index for each building is represented here with the final conclusion:

$$
\begin{aligned}
&\text{If Premise} \\
&(\text{"Premise variable a" is "Fuzzy set A" and "Premise variable b" is "Fuzzy set B"} \ldots) \\
&\text{then Consequence} \\
&(\text{"Consequence" is equal to "Number"}).
\end{aligned}
\tag{2}
$$

To compute the final conclusion, the single index, several steps are implemented:

1. Definition of rules based on fuzzy sets with a number for the conclusion (Equation (3));
2. Calculation of the membership functions for the fuzzy sets (Equation (4));
3. Calculation of the minimum membership function values per rule (Equation (5));
4. Calculation of the conclusion value per rule (Equation (6));
5. Computation of the final conclusion (Equation (7)).

$$
\begin{aligned}
&\text{Rule 1 with a consequence} = C1 \text{ (number)} \\
&\text{Rule 2 with a consequence} = C2 \text{ (number)} \\
&\cdots
\end{aligned}
\tag{3}
$$

$$\text{Membership function } \mu_A:X \to [0,1] \text{ where } \mu_A(x) \text{ is the membership value of x in A} \qquad (4)$$

$$\begin{aligned}
\text{Min(Rule 1)} &= \min(\mu_{A1}(x), \mu_{B1}(y), \dots) \\
\text{Min(Rule 2)} &= \min(\mu_{A2}(x), \mu_{B2}(y), \dots) \\
&\cdots
\end{aligned} \qquad (5)$$

$$\begin{aligned}
\text{Conclusion(Rule 1)} &= \text{Min(Rule 1)} \cdot C1 \\
\text{Conclusion(Rule 2)} &= \text{Min(Rule 2)} \cdot C2 \\
&\cdots
\end{aligned} \qquad (6)$$

$$\text{Final conclusion} = \sum \text{Conclusion(Rule i)} / \Sigma \text{Min(Rule i)}. \qquad (7)$$

A brief description of the implemented steps follows. For each geospatial parameter, two fuzzy logic sets and their corresponding membership functions are defined, based on their contribution to building vulnerability, as defined in introduction of Section 2.4. Linear functions are used, because they were also employed in the previous method, so later methods could be compared. Table 3 provides their names and spreadsheet expressions for their calculations, while Figure 5 illustrates them.

Table 3. Fuzzy sets and membership functions for the key geospatial parameters.

Geospatial Parameter	Linguistic Variable - Fuzzy Set -	Spreadsheet Expression for the Calculation of Fuzzy Membership Function Values
building's footprint area	Small building (SB)	=IF("area"<=15;1;IF("area"<45;(45-"area")/30;0))
	Large building (LB)	=IF("area"<=15;0;IF("area"<45;("area"-15)/30;1))
elevation above sea level	Low elevation (LE)	=IF("elevation"<=1;1;IF("elevation"<2;(2-"elevation");0))
	High elevation (HE)	=IF("elevation"<=1;0;IF("elevation"<2;("elevation"-1);1))
distance to coastline	Near to the sea (NS)	=IF("distance"<=30;1;IF("distance"<75;(75-"distance")/45;0))
	Far from the sea (FS)	=IF("distance"<=30;0;IF("distance"<75;("distance"-30)/45;1))

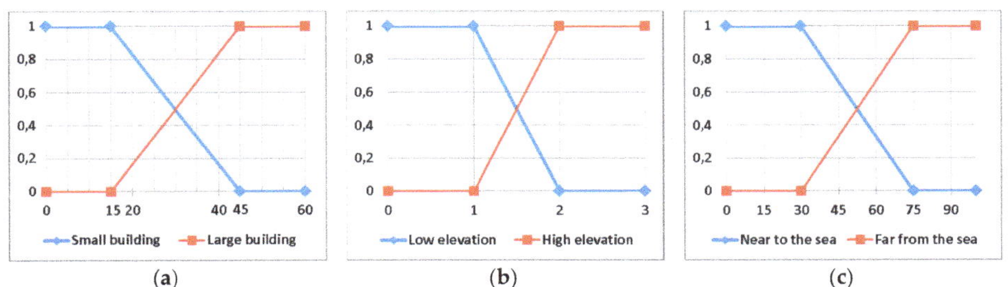

Figure 5. Graphs showing the fuzzy membership functions of key geospatial parameters: (**a**) small and large buildings for the building's footprint area; (**b**) low and high elevations for elevation above the sea level; (**c**) near and far from the sea for distance to coastline.

To define the rules, all combinations of fuzzy sets representing geospatial parameters should be considered. Therefore, there are 8 rules listed in Table 4. These rules' consequences (expressed as numbers and linguistic expressions) represent the vulnerability value that should be evaluated by an expert. The following steps include a calculation of the minimum membership function values per rule and a calculation of the conclusion value for the rule (a combination of Equations (5) and (6)); the spreadsheet expression is given in Table 4.

The final conclusion for each building is computed by Equation (7), and further rounded to the nearest integers of 1, 2, 3 4, and 5. Thus, the final conclusions represent the final building vulnerability indices expressed by linguistic expressions—the same ones used in the previous two methods.

Table 4. Rules with their linguistic expressions and spreadsheet expressions for the calculation of the rule conclusions.

No	Rule Premise	Consequence as a Fuzzy Singleton (Vulnerability Value)	Linguistic Expression of the Vulnerability Value	Rule Conclusion (Spreadsheet Expression)
1	If The building is small (SB), its elevation is low (LE), and it is near the sea (NS)	4	Medium high vulnerability	=MIN(SB;LE;NS)*4
2	If The building is small (SB), its elevation is low (LE), and it is far from the sea (FS)	3	Medium vulnerability	=MIN(SB;LE;FS)*3
3	If The building is small (SB), its elevation is high (HE), and it is near the sea (NS)	2	Medium low vulnerability	=MIN(SB;HE;NS)*2
4	If The building is small (SB), its elevation is high (HE), and it is far from the sea (FS)	1	Low vulnerability	=MIN(SB;HE;FS)*1
5	If The building is large (LB), its elevation is low (LE), and it is near the sea (NS)	5	High vulnerability	=MIN(LB;LE;NS)*5
6	If The building is large (LB), its elevation is low (LE), and it is far from the sea (FS)	4	Medium high vulnerability	=MIN(LB;LE;FS)*4
7	If The building is large (LB) its elevation is high (HE), and it is near the sea (NS)	3	Medium vulnerability	=MIN(LB;HE;NS)*3
8	If The building is large (LB), its elevation is high (HE), and it is far from the sea (FS)	2	Medium low vulnerability	=MIN(LB;HE;FS)*2

3. Results

Using the same values for the contribution of key geospatial parameters but different methods for their rankings and integration resulted in significant differences in the vulnerability indices assigned to the buildings. The results are summarised in Table 5 and illustrated in Figures 6–8.

Table 5. Numbers of buildings with vulnerability index 1, 2, 3, 4, and 5, calculated by the three methods.

Single Vulnerability Index	Index Method—Crisp Ranking	Index Method—Continuous Ranking	Fuzzy Logic Method
1	2	30	40
2	405	341	450
3	413	452	426
4	658	486	386
5	179	348	355
Total number of buildings	1657	1657	1657
Sum of single vulnerability indices for all buildings	5578	5752	5537

The sums of single vulnerability indices for all the buildings calculated by the three methods do not show significant differences; the greatest difference is 4% of the sum value. When considering particular buildings, there is a significant difference in their assessment using these three methods.

Examining the graph given in Figure 6, the greatest differences are found for vulnerability index 4 (among all three methods), while for vulnerability index 5, the Index method with crisp rankings has a significantly lower number of buildings. By introducing continuous ranking for the Index method, the results are closer to the Fuzzy logic method results.

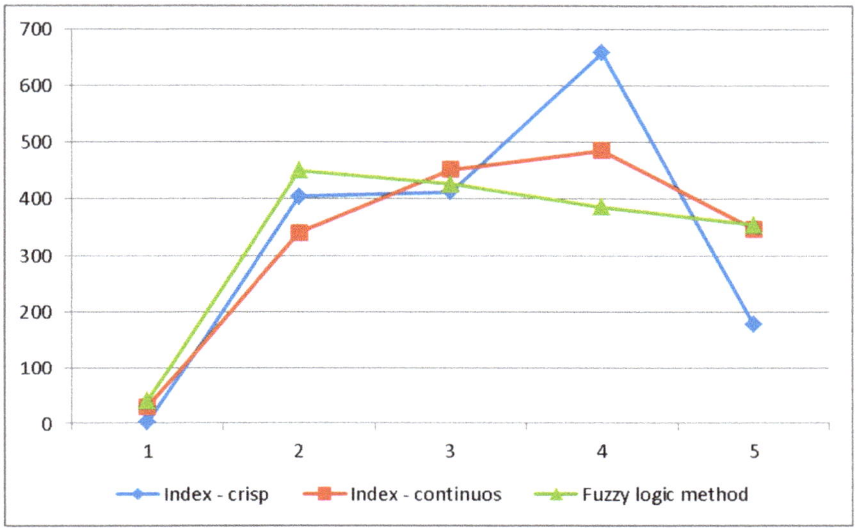

Figure 6. Graph showing the number of buildings with assigned single vulnerability indices of 1, 2, 3, 4, or 5, calculated by the three methods.

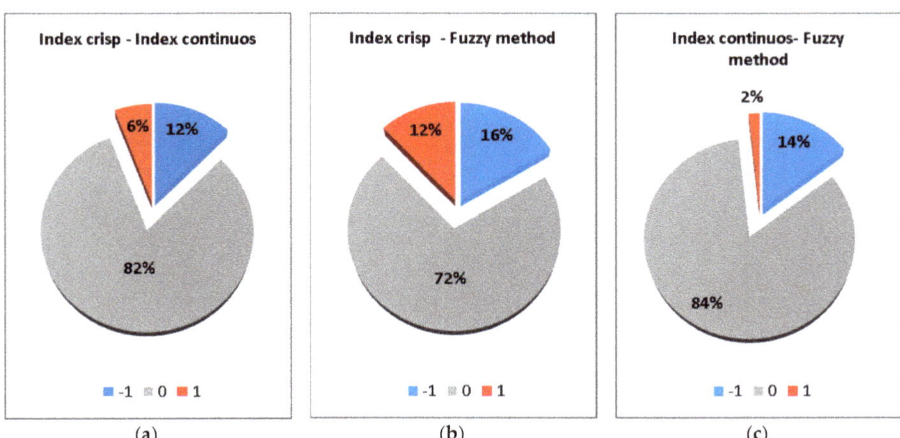

Figure 7. Graphs showing how many buildings have differences in their single vulnerability indexes calculated by the two methods (a blue colour represents a −1 score, red is a 1 score, and grey means no difference): (**a**) Index method with crisp ranking—Index method with continuous ranking; (**b**) Index method with crisp ranking—Fuzzy logic method; and (**c**) the Index method with continuous ranking—Fuzzy logic method.

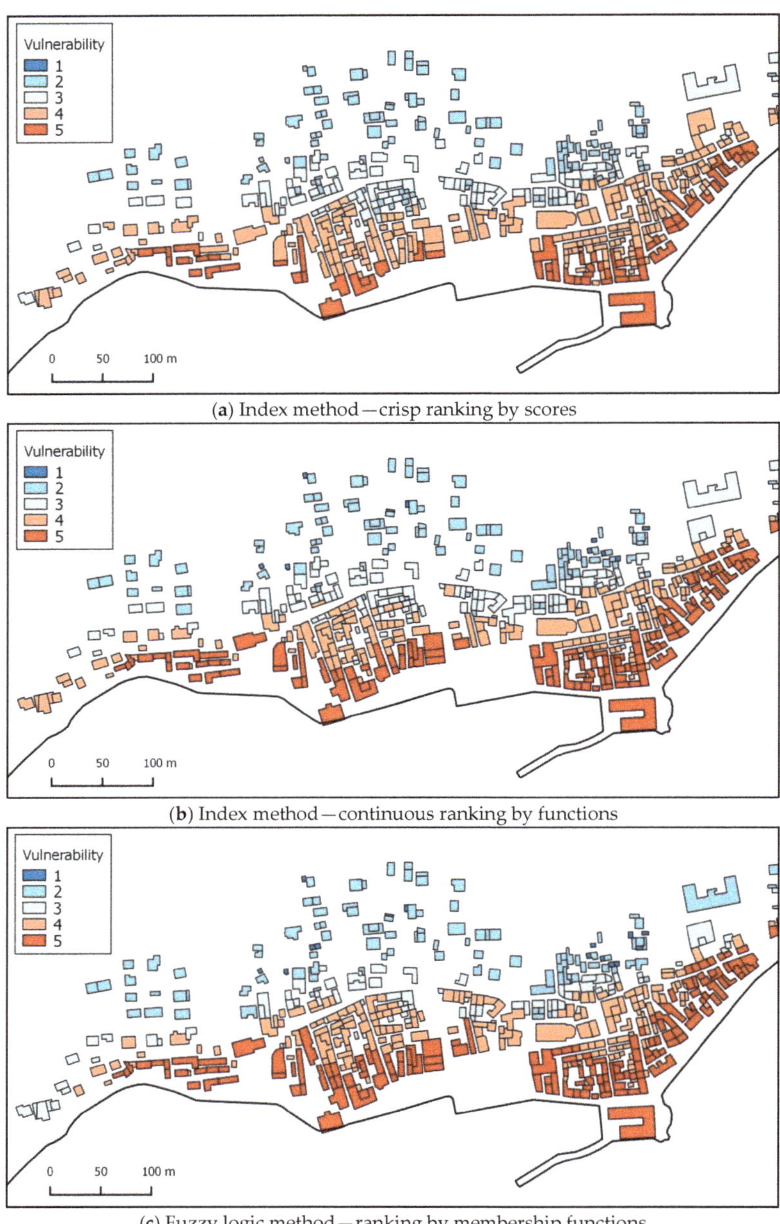

(**a**) Index method—crisp ranking by scores

(**b**) Index method—continuous ranking by functions

(**c**) Fuzzy logic method—ranking by membership functions

Figure 8. Maps showing the single vulnerability index for each building in the study area (an excerpt) calculated by the methods: (**a**) the Index method—crisp ranking by scores; (**b**) the Index method—continuous ranking by functions; and (**c**) the Fuzzy logic method—ranking by membership functions.

Examining the graphs given in Figure 7, the findings are as follows. Comparing the Index methods using crisp ranking with the methods using continuous ranking, there are significant differences in the building vulnerability assessment: 18% of buildings have a different vulnerability index, but there is no difference higher than 1 score. Comparing the Index method using crisp ranking and the Fuzzy logic method, the differences are even more significant: 28% of buildings have different vulnerability indexes, but, again, there is no difference higher than 1 score. Finally, comparing the Index method using continuous ranking and the Fuzzy logic method, there are still significant differences; however, these differences are a bit smaller: 16% of buildings have a different vulnerability index.

Finally, the spreadsheet table with all the calculated values is joined with the layer Building in QGIS, and the visualisations are performed. Figure 8 presents maps showing the single vulnerability indices for buildings calculated by the three methods: (a) the Index method—crisp ranking by scores; (b) the Index method—continuous ranking by functions; (c) and the Fuzzy logic method—ranking by membership functions. The highest vulnerability value (5) is coloured with red, and the colours change gradually to blue, which represents the lowest vulnerability value of 1. Summarising all the previous comparisons and observing the visualisations in Figure 8, one can conclude that the Index method with crisp ranking assigns more medium values (2, 3, and 4) and many fewer extreme values (1 and 5). The other two methods assigned significantly more buildings to the highest value of 5 (approximately 10% more), with more buildings assuming the lowest value because, in the Index method with crisp ranking, there are almost no values of 1.

4. Discussion

The above results show that there are significant differences in the vulnerability indices assigned to the buildings calculated by the various methods. Hence, the question arises: which method should one choose for a vulnerability assessment?

For the appropriate spatial units to be used, the relevant works clearly show that buildings are adequate units, as they are the key assets exposed. Moreover, socio–economic parameters are assigned to households, and can thus be easily assigned to building footprints. All three methods can use buildings or any other exposed assets as spatial units for vulnerability assessment.

Uncertainties in a vulnerability assessment sometimes originate in the data. Vagueness in the attribute values describing features cannot be avoided and causes uncertainties in the feature descriptions. The proposed Fuzzy logic method using the membership functions could model these uncertainties, while the other two methods cannot. Vagueness in the definition of boundaries causes uncertainties in the position and geometries of features. The solution proposed by the relevant work is based on Fuzzy Spatial Data Types whose implementation is complex and thus is not included in any of the proposed methods. Therefore, these types of uncertainties remain. There is also an uncertainty in the assessment caused by the aggregation of the key parameter values for each building. For example, the mean elevation and the shortest distance are calculated and assigned to each building, including their values within the building's footprint. Another source of uncertainty lies in the expert evaluation of the key parameters' contributions to vulnerability. These two uncertainties can be modelled by the membership functions; thus, the Fuzzy logic method can solve these uncertainties, while other the two methods cannot.

Crisp classifications, when applied to continuous phenomena, are unsatisfying. Both newly proposed methods overcome the pitfalls of existing Index-based methods, either by the introduction of functions or of fuzzy membership functions.

Finally, the methods are evaluated by how easily they can be applied by the local administration and accepted by all the participants. All three methods use widely available and open tools, as well as commonly available data for local administration. The Fuzzy logic method requires participant efforts in understanding the concepts of fuzzy set theory and fuzzy logic, while the other two methods are much simpler and more straightforward. However, the Fuzzy logic method is superior in supporting qualitative approaches and provides semantic values common to human perception. This is achieved by turning the quantitative parameters into linguistic variables such as "low" or "high" and by expressing the rules with linguistic expressions.

5. Conclusions

At the beginning of this research, we posed four questions that address the requirements of using an Index-based method for a household level analysis. Table 6 summarises a comparison of the existing method and the two newly proposed methods based on how they answer these questions.

Table 6. A comparison of the methods satisfying the requirements for the household level analysis.

	Question	Requirements of the Method for the Household Level Analysis	Index Method— Crisp Ranking	Index Method— Continuous Ranking	Fuzzy Logic Method
1.	What spatial units or tessellation types should be used for large-scale vulnerability assessment?	Assets exposed, e.g., buildings	yes	yes	yes
2.	How can one deal with uncertainties in vulnerability assessment?	Description uncertainties	no	no	yes
		Boundary uncertainties	no	no	no
		Key parameter value uncertainties	no	no	yes
		Expert evaluation uncertainties	no	no	yes
3.	How can one use Index and Indicator-based methods with crisp classification when key variables represent continuous phenomena?	Modelling continuous phenomena	no	yes	yes
4.	How could vulnerability be more easily accepted by local planners and accepted by society?	Method easily understood	yes	yes	no
		Using available tools and data	yes	yes	yes
		Supports qualitative approach	no	no	yes
		Semantics common to human perception	no	no	yes

Table 6 clearly shows that the Fuzzy logic method satisfies almost all the requirements and significantly improves the existing Index-based method, but it requires more effort from the human side. The local administration's staff and urban planners should become familiar with fuzzy set and fuzzy logic concepts, or they could become an obstacle to the method's application.

The Index method with continuous ranking introduces only one improvement: the modelling of continuous phenomena by functions. However, the results are significantly different from those obtained by the existing Index method with crisp ranking, and are instead closer to those of the Fuzzy logic method. The advantage of this method over the Fuzzy logic method is that the Index method is more simply understood by non-experts, and thus has more potential to be accepted and implemented by local administrations.

To determine which method to choose, there is always a trade-off between the comprehensiveness of the method and the resources needed for its implementation. This research showed that Index method with continuous ranking could be used as an alternative to the Fuzzy logic method. Thus, the Index method with continuous ranking is proposed for use by local administration as a simple and viable method that still improves vulnerability assessment at the household level.

Funding: This research was partially supported through the project KK.01.1.1.02.0027, a project co-financed by the Croatian Government and the European Union through the European Regional Development Fund–the Competitiveness and Cohesion Operational Programme.

Acknowledgments: The author is grateful to the editors and reviewers for their valuable comments on the manuscript.

Conflicts of Interest: The author declares no conflict of interest.

References

1. Ramieri, E.; Hartley, A.; Barbanti, A.; Santos, F.D.; Gomes, A.; Hilden, M.; Laihonen, P.; Marinova, N.; Santini, M. *Methods for Assessing Coastal Vulnerability to Climate Change*; Technical Paper 1; European Topic Centre on Climate Change Impacts, Vulnerability and Adaptation: Bologna, Italy, 2011.

2. Lavell, A.; Oppenheimer, M.; Diop, C.; Hess, J.; Lempert, R.; Li, J.; Muir-Wood, R.; Myeong, S. Climate change: New dimensions in disaster risk, exposure, vulnerability, and resilience. In *Managing the Risks of Extreme Events and Disasters to Advance Climate Change Adaptation*; A Special Report of Working Groups I and II of the Intergovernmental Panel on Climate Change, (IPCC); Field, C.B., Barros, V., Stocker, T.F., Qin, D., Dokken, D.J., Ebi, K.L., Mastrandrea, M.D., Mach, K.J., Plattner, G.-K., Allen, S.K., et al., Eds.; Cambridge University Press: Cambridge, UK; New York, NY, USA, 2012; pp. 25–64.

3. Koerth, J.; Vafeidis, A.; Hinkel, J. Household-Level Coastal Adaptation and Its Drivers: A Systematic Case Study Review. *Risk Anal.* **2016**, *37*, 629–646. [CrossRef] [PubMed]

4. Moret, W. *Vulnerability Assessment Methodologies: A Review of the Literature*; Family Health International (FHI 360): Durham, NC, USA, 2014.

5. Brooks, N. *Vulnerability, Risk and Adaptation: A Conceptual Framework*; Working Paper No. 38; Tyndall Centre for Climate Change Research: Norwich, UK, 2003.

6. Jean-Baptiste, N.; Kuhlicke, C.; Kunath, A.; Kabisch, S. *Review and Evaluation of Existing Vulnerability Indicators for Assessing Climate Related Vulnerability in Africa*; No. 07; UFZ-Bericht, Helmholtz-Zentrum für Umweltforschung: Leipzig, Germany, 2011.

7. Almås, A.J.; Hygen, H. Impacts of sea level rise towards 2100 on buildings in Norway. *Build. Res. Inf.* **2012**, *40*, 245–259. [CrossRef]

8. Gornitz, V.M. Vulnerability of the East coast, U.S.A. to future sea level rise. *J. Coast. Res.* **1990**, *9*, 201–237.

9. Gornitz, V.M. Global coastal hazards from future sea level rise. *Palaeogeogr, Palaeoclimatol. Palaeoecol. (Glob. Planet. Chang. Sect.)* **1991**, *89*, 379–398. [CrossRef]

10. Ozyurt, G. Vulnerability of Coastal Areas to Sea Level Rise: A Case of Study on Göksu Delta. Master's Thesis, Middle-East Technical University, Ankara, Turkey, January 2007. Available online: http://etd.lib.metu.edu.tr/upload/12608146/index.pdf (accessed on 8 October 2011).

11. Szlafsztein, C.; Sterr, H. A GIS-based vulnerability assessment of coastal natural hazards, State of Para, Brazil. *J. Coast. Conserv.* **2007**, *11*, 53–66. [CrossRef]

12. McLaughlin, S.; Cooper, J.A.G. A multi-scale coastal vulnerability index: A tool for coastal managers? *Environ. Hazards* **2010**, *9*, 233–248. [CrossRef]

13. Deduce Consortium. Available online: https://www.msp-platform.eu/practices/assessment-model-sustainable-development-european-coastal-zones (accessed on 15 January 2020).

14. Torresan, S.; Zabeo, A.; Rizzi, J.; Critto, A.; Pizzol, L.; Giove, S.; Marcomini, A. Risk assessment and decision support tools for the integrated evaluation of climate change on coastal zones. In Proceedings of the International Congress on Environmental Modelling and Software Modelling for Environment's Sake, Fifth Biennial Meeting, Ottawa, ON, Canada, 5–8 July 2010; Swayne, D.A., Wanhong, Y., Voinov, A.A., Rizzoli, A., Filatova, T., Eds.; International Environmental Modelling and Software Society: Manno, Switzerland, 2010.

15. Risk Assessment of Coastal Erosion: Part One. Available online: http://randd.defra.gov.uk/Document.aspx?Document=FD2324_7453_TRP.pdf (accessed on 15 January 2020).

16. DIVA. Available online: https://www.pik-potsdam.de/research/projects/projects-archive/favaia/diva (accessed on 15 January 2020).

17. Delft3D Modelling Suite. Available online: https://www.deltares.nl/en/software/delft3d-4-suite/ (accessed on 15 January 2020).

18. Miller, A.; Reiter, J.; Weiland, U. Assessment of urban vulnerability towards floods using an indicator-based approach—A case study for Santiago de Chile. *Nat. Hazards Earth Syst. Sci.* **2011**, *11*, 2107–2123. [CrossRef]

19. Kantamaneni, K.; Du, X.; Aher, S.; Singh, R. Building Blocks: A Quantitative Approach for Evaluating Coastal Vulnerability. *Water* **2017**, *9*, 905. [CrossRef]

20. Kim, Y.; Chung, E.S. An index-based robust decision making framework for watershed management in a changing climate. *Sci. Total. Environ.* **2014**, *473–474*, 88–102. [CrossRef]

21. Koroglu, A.; Ranasinghe, R.; Jiménez, J.; Dastgheib, A. Comparison of Coastal Vulnerability Index applications for Barcelona Province. *Ocean Coast. Manag.* **2019**, *178*, 104799. [CrossRef]

22. Basofi, A.; Fariza, A.; Dzulkarnain, M.R. Landslides susceptibility mapping using fuzzy logic: A case study in Ponorogo, East Java, Indonesia. In Proceedings of the 2016 International Conference on Data and Software Engineering (ICoDSE), Denpasar, Indonesia, 26–27 October 2016; pp. 1–7. [CrossRef]

23. Wardhana, M.; Sofwan, A.; Setiawan, I. Fuzzy Logic Method Design for Landslide Vulnerability. *E3S Web Conf.* **2019**, *125*, 03004. [CrossRef]

24. Sadrykia, M.; Delavar, M.; Mehdi, Z.A. GIS-Based Fuzzy Decision Making Model for Seismic Vulnerability Assessment in Areas with Incomplete Data. *ISPRS Int. J. Geo-Inf.* **2017**, *6*, 119. [CrossRef]

25. Jadidi, M.; Mostafavi, M.A.; Bédard, Y.; Kyarash, S. Spatial Representation of Coastal Risk: A Fuzzy Approach to Deal with Uncertainty. *ISPRS Int. J. Geo-Inf.* **2014**, *3*, 1077–1100. [CrossRef]

26. Rashetnia, S. Flood Vulnerability Assessment by Applying a Fuzzy Logic Method: A Case Study from Melbourne. Master's Thesis, Victoria University, Melbourne, Australia, August 2016.

27. Galindo, J.; Schneider, M. Fuzzy Spatial Data Types for Spatial Uncertainty Management in Databases. In *Handbook of Research on Fuzzy Information Processing in Databases*, 1st ed.; Galindo, J., Ed.; IGI Global: Hershey, PA, USA, 2008; pp. 154–196. [CrossRef]

28. Chu, T.C. Selecting plant location via a Fuzzy TOPSIS approach. *Int. J. Adv. Manuf. Technol.* **2002**, *20*, 859–864. [CrossRef]

29. Yong, D. Plant location selection based on fuzzy TOPSIS. *Int. J. Adv. Manuf. Technol.* **2006**, *28*, 839–844. [CrossRef]

30. Jun, K.S.; Chung, E.S.; Kim, Y.G.; Kim, Y. A fuzzy multicriteria decision approach to flood risk vulnerability in South Korea by considering climate change impacts. *Expert Syst. Appl.* **2013**, *40*, 1003–1013. [CrossRef]

31. Kim, Y.; Chung, E.S.; Jun, S.M.; Kim, S.U. Prioritizing the best sites for treated wastewater use in an urban watershed using Fuzzy TOPSIS. *Resour. Conserv. Recycl.* **2013**, *73*, 23–32. [CrossRef]

32. Jara, J.; Acevedo-Crespo, R. Crisp Classifiers vs. Fuzzy Classifiers: A Statistical Study. *Comput. Vision* **2009**, *5495*, 440–447. [CrossRef]

33. Vadiati, M.; Asghar, M.A.; Nakhaei, M.; Adamowski, J.; Akbarzadeh, A.H. A fuzzy-logic based decision-making approach for identification of groundwater quality based on groundwater quality indices. *J. Environ. Manag.* **2016**, *184*. [CrossRef]

34. Valentini, E.; Nguyen Xuan, A.; Filipponi, F.; Taramelli, A. Coastal vulnerability assessment using Fuzzy Logic and Bayesian Belief Network approaches. In *Geophysical Research Abstracts 201, 19, EGU2017-18063*; European Geosciences Union General Assembly: Vienna, Austria, 2017.

35. Akter, T.; Simonovic, S. Aggregation of Fuzzy Views of a Large Number of Stakeholders for Multi-Objective Flood Management Decision-Making. *J. Environ. Manag.* **2005**, *7*, 133–143. [CrossRef]

36. Lee, G.; Jun, K.S.; Eun-Sung, C. Group decision-making approach for flood vulnerability identification using the fuzzy VIKOR method. *Nat. Hazards Earth Syst. Sci. Discuss.* **2014**, *2*. [CrossRef]

37. Shan, X.; Wen, J.; Zhang, M.; Wang, L.; Ke, Q.; Li, W.; Du, S.; Shi, Y.; Chen, K.; Liao, B.; et al. Scenario-Based Extreme Flood Risk of Residential Buildings and Household Properties in Shanghai. *Sustainability* **2019**, *11*, 3202. [CrossRef]

38. Jadidi, M.; Mostafavi, M.A.; Bédard, Y.; Long, B.; Grenier, E. Using geospatial business intelligence paradigm to design a multidimensional conceptual model for efficient coastal erosion risk assessment. *J. Coast. Conserv.* **2013**, *17*, 527–543. [CrossRef]

39. Bermúdez, M.; Zischg, A.P. Sensitivity of flood loss estimates to building representation and flow depth attribution methods in micro-scale flood modelling. *Nat. Hazards* **2018**, *92*, 1633–1648. [CrossRef]

40. Hatzikyriakou, A.; Lin, N. Assessing the Vulnerability of Structures and Residential Communities to Storm Surge: An Analysis of Flood Impact during Hurricane Sandy. *Front. Built Environ.* **2018**, *4*. [CrossRef]

41. Swart, R.; Fons, J.; Geertsema, W.; Hove, L.V.; Jacobs, C. *Urban Vulnerability Indicators. A Joint Report of ETC-CCA and ETC-SIA. ETC CCA/ETC/SIA*; Technical Report 01; ETC CCA: Bologna, Italy, 2012.

42. Coastal Plan for the Šibenik-Knin County (2015, PAP/RAC). Available online: http://iczmplatform.org/storage/documents/pEoju2FqfXjzPoYBLsKZiD3o6ONBXxJ44RTWFt7P.pdf (accessed on 15 January 2020).

43. Andričević, R.; Knezić, S.; Vranješ, M.; Baučić, M.; Jajac, N. Report on Initial Flood Vulnerability Assessment in the Sava River Basin, Pilot Project on Climate Change: Building the Link between Flood Risk Management Planning and Climate Change Assessment in the Sava River Basin, the International Sava River Basin Commission, 2013. Available online: https://www.savacommission.org/project_detail/17/1 (accessed on 15 January 2020).

44. Margeta, J.; Vilibić, I.; Jakl, Z.; Marasović, K.; Petrić, L.; Mandić, A.; Grgić, A.; Bartulović, H.; Baučić, M. *Draft version: Coastal Action Plan for the City of Kaštela*; Technical Report; JU RERA: Split, Croatia, 2019.

45. Baučić, M.; Ivić, M.; Jovanović, N.; Bačić, S. Vulnerability analysis for the integrated coastal zone management plan of the City of Kaštela in Croatia. *ISPRS–Int. Arch. Photogramm. Remote Sens. Spat. Inf. Sci.* **2019**, *XLII-3/W8*, 59–63. [CrossRef]

46. Begg, S.H.; Welsh, M.B.; Bratvold, R.B. Uncertainty vs. Variability: What's the Difference and Why is it Important? In Proceedings of the Society of Petroleum Engineers Hydrocarbon Economics and Evaluation Symposium, Houston, TX, USA, 19–20 May 2014. [CrossRef]

47. Fisher, P.; Comber, A.; Wadsworth, R. Approaches to Uncertainty in Spatial Data. In *Fundamentals of Spatial Data Quality*, 1st ed.; Devillers, R., Jeansoulin, R., Eds.; ISTE Ltd.: London, UK, 2010; pp. 43–59. [CrossRef]

48. Zadeh, L.A. Fuzzy sets. *Inf. Control* **1965**, *8*, 338–353. [CrossRef]

49. Proverbs, D.G.; Soetanto, R. *Flood Damaged Property, A Guide to Repair*, 1st ed.; Blackwell Publishing Ltd: Hoboken, NJ, USA, 2004.

50. Open Digital Elevation Model (OpenDEM). Available online: https://www.opendem.info/index.html (accessed on 15 January 2020).

51. Šimek, K.; Medak, D.; Medved, I. Analiza visinske točnosti službenoga vektorskoga digitalnoga modela reljefa Republike Hrvatske dobivenog fotogrametrijskom restitucijom. *Geod. List* **2018**, *3*, 217–230.

52. Brovelli, M.; Zamboni, G.A. New Method for the Assessment of Spatial Accuracy and Completeness of OpenStreetMap Building Footprints. *ISPRS Int. J. Geo-Inf.* **2018**, *7*, 289. [CrossRef]

53. QGIS. A Free and Open Source Geographic Information System. Available online: https://qgis.org/en/site/ (accessed on 15 January 2020).

International Journal of
Geo-Information

Article

Effectiveness of Sentinel-2 in Multi-Temporal Post-Fire Monitoring When Compared with UAV Imagery

Luís Pádua [1,2,*], **Nathalie Guimarães** [1], **Telmo Adão** [1,2], **António Sousa** [1,2], **Emanuel Peres** [1,2] and **Joaquim J. Sousa** [1,2]

[1] Engineering Department, School of Science and Technology, University of Trás-os-Montes e Alto Douro, 5000-801 Vila Real, Portugal; nsguimaraes@utad.pt (N.G.); telmoadao@utad.pt (T.A.); amrs@utad.pt (A.S.); eperes@utad.pt (E.P.); jjsousa@utad.pt (J.J.S.)

[2] Centre for Robotics in Industry and Intelligent Systems (CRIIS), INESC Technology and Science (INESC-TEC), 4200-465 Porto, Portugal

* Correspondence: luispadua@utad.pt; Tel.: +351-259-350-762 (ext. 4762)

Received: 30 January 2020; Accepted: 3 April 2020; Published: 7 April 2020

Abstract: Unmanned aerial vehicles (UAVs) have become popular in recent years and are now used in a wide variety of applications. This is the logical result of certain technological developments that occurred over the last two decades, allowing UAVs to be equipped with different types of sensors that can provide high-resolution data at relatively low prices. However, despite the success and extraordinary results achieved by the use of UAVs, traditional remote sensing platforms such as satellites continue to develop as well. Nowadays, satellites use sophisticated sensors providing data with increasingly improving spatial, temporal and radiometric resolutions. This is the case for the Sentinel-2 observation mission from the Copernicus Programme, which systematically acquires optical imagery at high spatial resolutions, with a revisiting period of five days. It therefore makes sense to think that, in some applications, satellite data may be used instead of UAV data, with all the associated benefits (extended coverage without the need to visit the area). In this study, Sentinel-2 time series data performances were evaluated in comparison with high-resolution UAV-based data, in an area affected by a fire, in 2017. Given the 10-m resolution of Sentinel-2 images, different spatial resolutions of the UAV-based data (0.25, 5 and 10 m) were used and compared to determine their similarities. The achieved results demonstrate the effectiveness of satellite data for post-fire monitoring, even at a local scale, as more cost-effective than UAV data. The Sentinel-2 results present a similar behavior to the UAV-based data for assessing burned areas.

Keywords: post-fire management; forest regeneration; fire severity mapping; multispectral imagery; Sentinel-2A; unmanned aerial vehicles; Parrot SEQUOIA

---

1. Introduction

In recent years, forest fires (i.e., large and destructive fires that spread over a forest or area of woodland) have received increasing attention due to their effects on climate change and ecosystems. Forest fires occur regularly, vary in scale and impacts and are inherent to terrestrial ecosystems [1]. Weather, topography and fuel are the three major components that define the fire environment and are directly related with the evolution of land use [2]. Portugal is one of the southern European countries most affected by forest fires, but it is also affected by rural fires [3]. In other words, not only do fires over forests affect the country, but the combination of environmental factors and human settlement may also cause harm to people or damage property or the environment [4]. Several factors contribute to the country being so severely affected: the Mediterranean climate, which benefits fuel accumulation and dryness along with the existence of flammable vegetation types; high ignition density; poor fire-suppression capabilities; and institutional instability [5]. Thus, forest fire impacts are attracting more and more attention not only from the scientific community, but also from public entities worldwide [5]. In the Portuguese case, this awareness is increasing, especially in the north and in the center of the country [6].

In this context, remote sensing platforms are being used as a capable tool for mapping burned areas, evaluating the characteristics of active fires and characterizing post-fire ecological effects and regeneration [7]. In the past decade, the use of unmanned aerial vehicles (UAVs) has increased for agroforestry applications [8] and are now being used for forest fire prevention [9], canopy fuel estimation [10], fire monitoring [11,12] and to support firefighting operations [13]. Likewise, studies using UAV-based imagery in post-fire monitoring have been concerned with surveying [14], calibrating satellite-based burn severity indices [15], assessing post-fire vegetation recovery [16], mapping fire severity [17,18], studying forest recovery dynamics [19] and sapling identification [20]. Despite being a cost-effective and a very versatile platform for remote sensed data acquisition that is capable of carrying a wide set of sensors, its usage in surveying big areas can be constrained due to legal [21] and technological limitations such as its autonomy and payload capacity [8]. On the other hand, traditional remote sensing platforms such as satellites continue to be widely used to obtain data with increasingly improved spatial, temporal and radiometric resolutions. Satellites still offer a quick way to evaluate forest regeneration in post-fire areas. However, lower spatial resolutions (compared with UAV data) often mean that satellites are used for studies only at regional or national scales [22–26]. The Copernicus Programme, from the European Union's Earth Observation Programme, was created with the goal to achieve a global, continuous, autonomous, high-quality, wide-range Earth observation capacity. The different satellite missions belonging to this program make it possible to obtain accurate, timely and easily accessible information to improve the management of the environment, as well as to understand and mitigate the effects of climate change and ensure civil security. Therefore, access to medium- and high-resolution satellite data with a high temporal resolution are accessible for free [27], namely, the Sentinel-2 Multispectral Instrument (MSI) [28]. A wide range of spectral bands are available from visible to shortwave infrared (SWIR) which allows, in a post-fire monitoring context, severity determination of fire disturbances along with multi-temporal monitoring for burnt areas. This type of data is ideal for monitoring fire disturbances in Mediterranean regions that affect several crops and have extents ranging from some hectares to several square kilometers [29]. In this specific context, Sentinel-2 MSI data were used for exploring spectral indices of burn severity discrimination [30–34], as well as to assess burn severity in combination with Landsat data [35,36]. They were also used to take into account the available multi-temporal data in order to evaluate burned areas at a national level [37] and to assess post-fire vegetation recovery mapping of an island [38].

In this study, we evaluated an area that was severely affected by a fire disturbance in 2017 with an estimated extent greater than 300 ha. The area is located in north-eastern Portugal, and forested areas composed of maritime pine (*Pinus pinaster*) were significantly affected along with houses, wood storage buildings, agricultural structures and vehicles. This was therefore a fire that could be considered small, its analysis and monitoring could be possible to carry out using aerial high-resolution data acquired by a UAV. Every year, thousands of fires similar to this occur in Portugal, covering the north and center of the country in particular with small patches of burnt areas.

To assess the effectiveness of satellite data in studying this specific type of area, Sentinel-2 MSI data were used to characterize the area before the fire disturbance, allowing an assessment of the fire's severity and multi-temporal analysis to be performed (2017–2019). Moreover, to compare the spatial information provided by the Sentinel-2 MSI (10-m spatial resolution), a UAV flight campaign was carried in part of the study area to acquire multispectral data with a very-high resolution. This is precisely the central question of this study: what is the potential use of new generation free-access satellite images (Sentinel-2) to monitor small-scale burnt areas. To the best of our knowledge, this is the first study that uses freely available satellite data to analyze a burnt area of relatively small dimensions and conclude that the results were in line with those obtained by high-resolution data acquired by a UAV. Although more studies are needed that cover different areas with different complexities and different vegetation covers, this study allowed us to conclude that satellite data have great potential, in certain cases, to replace high aerial-resolution data acquired by UAVs. This would allow analyzing post-fire areas (even small ones) at the national level, representing considerable savings in time and money.

2. Materials and Methods

2.1. Study Area

The study area, highlighted in Figure 1, is located in the north-eastern region of Portugal within the municipality of Sabrosa (41°20′40.4″ N, 7°36′04.5″ W), near the villages of Parada do Pinhão and Vilarinho de Parada. This area was severely affected by a wildfire that began at 12:59 p.m. on 13 August 2017 and was reported as extinguished at 03:16 a.m. on 14 August 2017 [39]. The area is characterized by a warm and temperate climate, an average annual temperature of 13.1 °C and an annual precipitation averaging 1139 mm. July and August are the months with the highest mean temperatures (21 °C) and lower precipitation (28 mm). This area was selected due to its easy accessibility and representativeness, since the species in the area are common for the region. It is mostly populated by maritime pine, deciduous species such as *Quercus pyrenaica* and *Castanea sativa* Mill. and some riparian species, shrubland communities and parcels used for agriculture and silviculture purposes. Moreover, the burned area was greater than 100 ha, which fit the majority of the fire events (93%) that occurred in Portugal during 2017 [6].

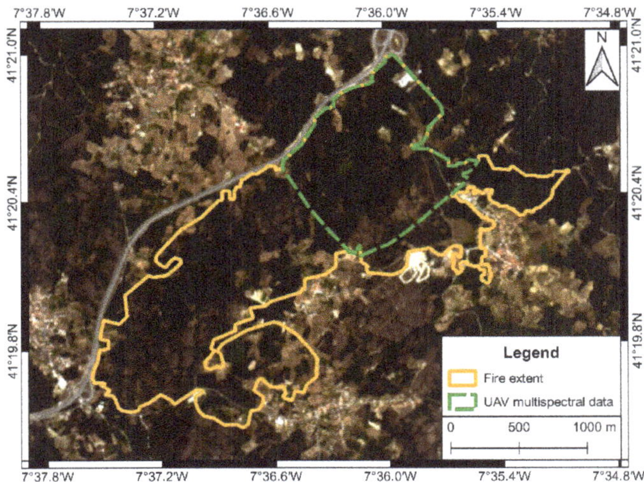

Figure 1. Overview of the study area along with the fire extent and area surveyed by the unmanned aerial vehicle (UAV).

2.2. Remote Sensing Dataset

The satellite imagery data used in this study were acquired by the Sentinel-2 MSI. Spectral data products provided by MSI ranged from the visible to the shortwave infrared (SWIR) parts of the electromagnetic spectrum. In total, there were 13 available spectral bands (B) at different spatial resolutions: (1) at 10 m—B2 (490 nm), B3 (560 nm), B4 (665 nm) and B8 (842 nm); (2) at 20 m—B5 (705 nm), B6 (740 nm), B7 (783 nm), B8a (865 nm), B11 (1610 nm) and B12 (2190 nm); and (3) at 60 m—B1 (443 nm), B9 (940 nm) and B10 (1375 nm) [28]. Data were obtained from the Copernicus Open Access

Hub with an absence of clouds over the study area from June 2017 to October 2019. These epochs were selected due to being related to the last available period before the fire disturbance (June, July and August 2017), including the first cloud and smoke-free post-fire data (September 2017). The imagery was atmospherically corrected using the Sen2Cor [40].

Regarding UAV data, the senseFly eBee (senseFly SA, Lausanne, Switzerland) was used to acquire both RGB and multispectral imagery. A Canon IXUS 127 HS sensor with 16.1 MP resolution was used for RGB imagery acquisition, and the Parrot SEQUOIA sensor was used for multispectral imagery acquisition. The multispectral sensor comprised a four-camera array with 1.2 MP resolution acquiring green (530–570 nm), red (640–680 nm), red edge (730–740 nm) and near infrared (NIR) (770–810 nm) imagery. Its radiometric calibration was performed using a target prior to the flight. Two flights with the same mission plan (one per sensor) were performed on 11 July 2019. The RGB flight was performed at a 425-m height, covering an area of 230 ha, with a spatial resolution of 0.12 m. The imagery overlap was 80% front and 60% side, for a total acquisition of 91 georeferenced images (related to a ground system of geographic coordinates) distributed through eight strips (approximately 11 images per strip). As for the multispectral flight, it was carried out at a 215-m height, covering approximately 150 ha, with a spatial resolution of approximately 0.25 m; it had an 80% front overlap and 60% side overlap, for a total acquisition of 260 images per spectral band (12 strips with approximately 22 images per strip).

A pre-processing of UAV-based imagery is required before it is ready for use. Thus, Pix4Dmapper Pro version 4.4.12 (Pix4D SA, Lausanne, Switzerland) was used for the photogrammetric processing of the UAV imagery, and common tie points were identified in the provided imagery according to their geolocation and internal and external camera parameters. This enabled the computation of dense 3D point clouds that were further interpolated using inverse distance weighting (IDW) to obtain the following orthorectified outcomes: an orthophoto mosaic from the RGB imagery, digital elevation models (DEMs) and four radiometric bands from the multispectral imagery that could then be used for the computation of vegetation indices. DEMs were not used in the scope of this study, and the orthophoto mosaic was used for visual inspection only.

2.3. Data Processing and Analysis

Both satellite and UAV multispectral datasets were used to compute vegetation indices. Sentinel-2-based vegetation indices were used to assess the fire severity and to perform the post-fire multi-temporal analysis of the study area. Similar vegetation indices were computed using UAV data for a single epoch, allowing a comparison of both sets of results.

2.3.1. Computation of Spectral Indices

The satellite data were used to compute the normalized burn ration (NBR) [41] as in Equation (1). This index relates to vegetation moisture content by combining the NIR (B8) and SWIR (B12) parts of the electromagnetic spectrum [42], and is generally accepted as a standard spectral index to assess burn severity [41,43].

$$NBR = \frac{NIR - SWIR}{NIR + SWIR} \tag{1}$$

Moreover, the normalized difference vegetation index (NDVI) [44] was calculated using a NIR band (B8) and a RED band (B4) from Sentinel-2 MSI data. NIR and RED bands from the UAV-based multispectral data were also used to compute the equivalent index, as in Equation (2). NDVI is widely used to analyze the vegetation condition in different contexts [8].

$$NDVI = \frac{NIR - RED}{NIR + RED} \tag{2}$$

2.3.2. Post-Fire Multi-Temporal Analysis

The multi-temporal analysis performed in this study relied on the time series data provided by the Sentinel-2 MSI. From the available data, a set of four epochs was selected for each year (2017 to 2019), corresponding each one to the months of June, July, August and September, with the dates of the selected data presented in Table 1. This period was selected (June, July and August 2017) in order to include data prior to the fire disturbance, along with the same months in following available years (2018 and 2019). Some data outside these periods were affected by clouds and had to be discarded. Moreover, it was decided to not consider any data from October to May in order to avoid false assumptions from the natural seasonal behavior of the species in the study area (e.g., the absence of leaves in deciduous tree species in the winter time, and potential interference of undergrowth vegetation in winter and spring time). The selected months assured that the trees were fully developed and that undergrowth vegetation interference was minimal [45].

Table 1. Days corresponding to the Sentinel-2 data selected for multi-temporal analysis. June, July and August 2017 correspond to data before the fire disturbance.

Year	Month			
	June	July	August	September
2017	4	14	13	22
2018	24	29	23	12
2019	29	19	13	12

The difference normalized burn ration (dNBR), calculated by subtracting the post-fire raster data from the pre-fire raster as in Equation (3), was used to perform the burn severity level classification as proposed by the United States Geological Survey (USGS) [46,47], enabling an understanding not only the severity of the burned areas, but also of unburned areas within the study region. Pre- and post-fire NBRs were the NBR of a date before and after the fire disturbance, respectively. In burned areas, the NBR showed higher values before the fire and lower values after the fire. The dNBR was the difference between the NBRs of both epochs: positive values represented areas with a higher fire severity, while values close to or lower than zero represented unburned areas and/or vegetation regrowth. For each classified severity level, the mean NDVI value was calculated per analyzed month. The mean NDVI value was also estimated for the whole burned area.

$$dNBR = PrefireNBR - PostfireNBR \tag{3}$$

To evaluate the post-fire recovery, a similar analysis was performed using the difference NDVI (dNDVI) by subtracting the NDVI of first post-fire (September 2017) from the NDVI values from each analyzed month from 2018 and 2019. This way, positive values represented an increase in the NDVI and, consequently, a potential recovery zone, while the inverse was true for values close to or less than zero.

The data analysis was carried in the opensource geographical information system (GIS) QGIS (version 3.4.12-Madeira) and functions from the Geographic Resources Analysis Support System (GRASS GIS) [48] and from the System for Automated Geoscientific Analyses (SAGA GIS) [49] were also used.

2.3.3. Sentinel-2 MSI and UAV Comparison

The Sentinel-2 MSI data acquired on 9 July 2019 were compared to the UAV-based multispectral imagery (two days difference). The NDVI maps produced from both datasets were compared. The UAV-based NDVI at its original spatial resolution (0.25 m), its resampling to half the resolution and its resampling to same resolution as the Sentinel-2 MSI (5 and 10 m, respectively) were used for this comparison. A total of 116 ha (~35%) of the burned area (Figure 1) was evaluated. This is precisely the most complex area, containing a greater variety of tree species, agricultural fields and infrastructure. The resampling of the UAV NDVI was performed using the "r.resamp.stats" function from GRASS GIS in QGIS, by specifying the grid cell sizes (5 × 5 m and 10 × 10 m) and assigning the aggregated mean values to each cell. The correlation among the different NDVIs (UAV-based and satellite-based) was performed using the "r.covar" function from GRASS GIS.

Moreover, the geospatial variability of the Sentinel-2 NDVI was compared with the UAV NDVI at the three different spatial resolutions. The mean values of each evaluated NDVI were quantified in a 50 × 50 m grid. The size of this grid, representing five times the Sentinel-2 resolution, was selected to smooth the transition zones of vegetation cover. Then, the local bivariate Moran's index (MI) [50] and the bivariate local indicators of spatial association (BILISA) [51] were applied as in Anselin [52] to assess the spatial relationship between the NDVIs computed from both datasets. The local bivariate MI was used to assess the correlation between a defined variable (satellite NDVI) and a different variable in the nearby areas (UAV NDVI). BILISA was used to measure the local spatial correlation, forming maps of clusters with similar behaviors and enabling an assessment of their spatial variabilities and dispersion. These cluster maps were divided into four classes based on the correlation of a value with its neighborhood: high–high (HH); low–low (LL); high–low (HL); and low–high (LH). This analysis was made using GeoDa software (version 1.14.0) [53]. The required weights map was defined using an eight-connectivity approach (3 × 3 matrix) and 999 random permutations were used in the BILISA execution.

3. Results

3.1. Sentinel-2 Post-Fire Monitoring

The fire severity map calculated using the dNBI from the pre-fire NBI (August 2017) and the first post-fire NBI (September 2017) are presented in Figure 2. From the 361 ha representing the study area, 42% (151 ha) presented a high severity, 44% (160 ha) showed a moderate severity and 38 ha (11%) presented a low severity. Only 3% of the area (12 ha) was estimated not to have been affected by the fire disturbance. A visual inspection of these areas allowed us to conclude that unburned and low-severity areas represented infrastructures, or corresponded to bare soil or fields used for agriculture along with some tree stands. Moderate severity areas included shrubland communities, agriculture terrains and trees, while high-severity areas mostly included highly density forest stands.

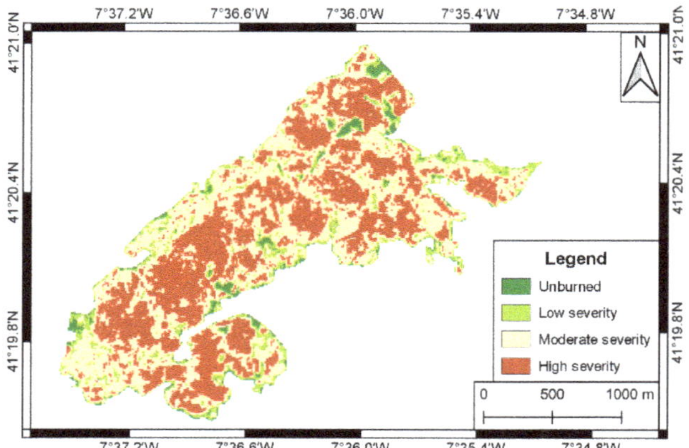

Figure 2. Fire severity classification of the study area.

The Sentinel-2 multi-temporal data enabled us to characterize the study area throughout the analyzed period. Figure 3 presents the pre- and post-fire NDVI (August and September 2017,

Figure 3a,b) and the NDVI from September of the two subsequent years (2018 and 2019, Figure 3c,d). The fire disturbance is clearly observable from the NDVI data and some forestry recovery is noticeable in the north, north-eastern and south-western parts of the study area. This is especially distinguishable in 2019 (Figure 3g).

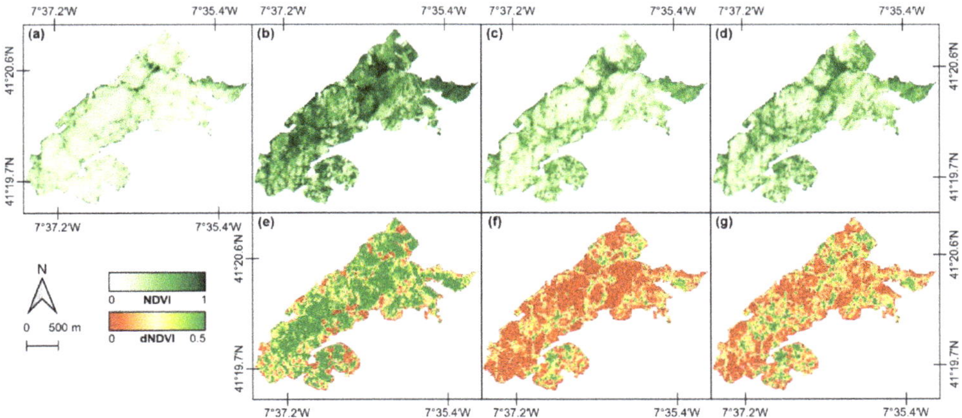

Figure 3. Normalized difference vegetation index (NDVI) of the study area in: (**a**) September 2017, (**b**) August 2017, (**c**) September 2018 and (**d**) September 2019. The difference NDVI (dNDVI) compared to September 2017 in (**e**) August 2017, (**f**) August 2018 and (**g**) August 2019.

The mean NDVI value was extracted for each severity level and unburned area for the months of June, July, August and September during 2017–2019, as well as for the whole area affected by the fire. Figure 4 presents these results. When analyzing the values obtained from the whole area (Figure 4a), the decline of NDVI values (−56%) after the fire disturbance (August to September 2017) is clearly noticeable. From September 2017 to June 2018, a growth of 52% in the mean NDVI value was verified, while in the homologous period in 2019 the growth was 33%. When separately analyzing each year, the values declined each month, with less noticeable results from August to September.

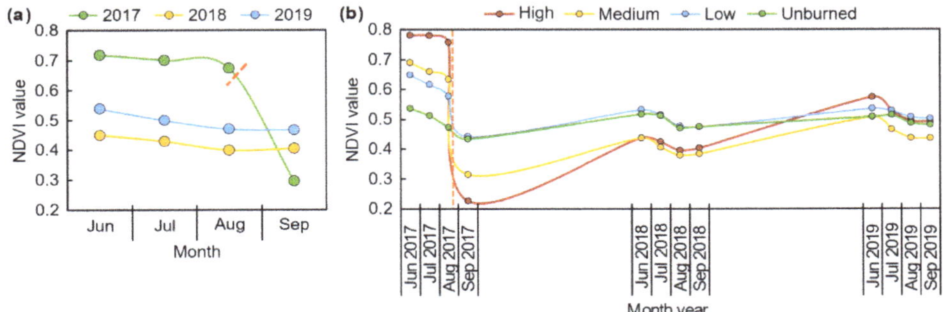

Figure 4. Mean values of the normalized difference vegetation index (NDVI) of the study area (**a**) and for each severity level (**b**) in the months of June, July, August and September of 2017, 2018 and 2019. The orange dashed line marks the fire disturbance date.

This tendency is reflected when observing the mean NDVI values per severity level (Figure 4b). The mean NDVI value of the unburned area was relatively constant, with a standard deviation of 0.03. Similarly, the area classified as low severity presented a standard deviation of 0.06. On the other hand, high-severity areas presented higher post-fire increases (with a standard deviation of 0.06 considering 2018 and 2019 values, and 0.17 overall), and the mean NDVI value presented growths of 93% from September 2017 to June 2018 and 42% from September 2018 to June 2019. For the moderate severity areas, these increases were 39% for September 2017 to June 2018, and 32% for September 2018 to June 2019, with a standard deviation of 0.04 (0.12 for the whole period). By comparing June 2018 to June 2019, the mean NDVI values for the high-, moderate- and low-severity areas and unburned areas presented variations of 32%, 16%, 1% and −2%, respectively.

When analyzing the post-fire dNDVIs (Figure 5) relating the differences in the first post-fire data (September 2017), a similar trend was observed. By analyzing the mean differences per year, an overall mean difference of 0.12 was verified in 2018, while in 2019 this difference was 0.20. In both years the same trend was verified, with the higher differences verified in areas with high severity, followed by moderate-severity areas. Both unburned and low-severity areas presented lower differences, with a mean difference of 0.06 for the two classes in 2018, an increase to 0.08 in 2019 for the low-severity areas and the same value maintained for the unburned area. The values declined from June to August and remained similar in September. When comparing July 2018 to July 2019, an overall increase of 0.09 was verified in the mean dNDVI values, representing increases of 0.14, 0.07, 0 and −0.01 for high, moderate, low-severity and unburned areas, respectively. A visual representation of the pre- and post-fire dNDVIs for the two subsequent years is presented in Figure 3e–g.

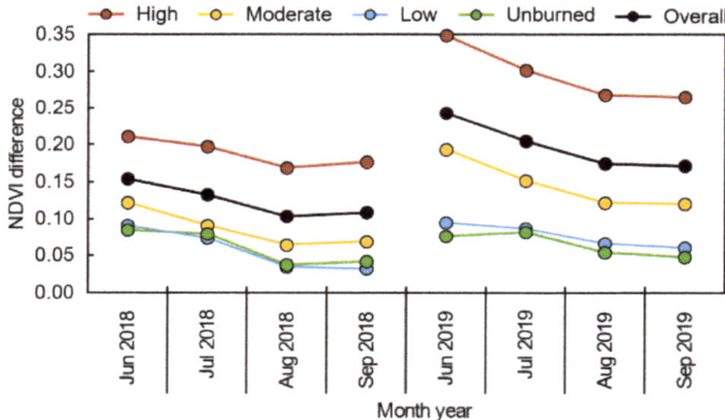

Figure 5. Mean values of the difference normalized difference vegetation index (dNDVI) of the study area for each severity level in the months of June, July, August and September during 2018 and 2019.

3.2. Comparison of UAV-Based and Sentinel-2 MSI Data

As mentioned in Section 2.3.3., the UAV-based multispectral data covered 116 ha of the study area. This was used to perform a comparison between the Sentinel-2 NDVI and the UAV-based NDVI at different spatial resolutions (Figure 6). The statistics of the different spatial resolutions of the UAV NDVI (Table 2) were similar in their mean values, while the minimum, maximum and standard deviation values tended to be greater for higher spatial resolutions. In regards to the NDVI computed from the Sentinel-2 dataset, a small difference was verified for the mean value, while the minimum, maximum and standard deviation values were similar to the UAV NDVI at a 10-m spatial resolution (Figure 6c).

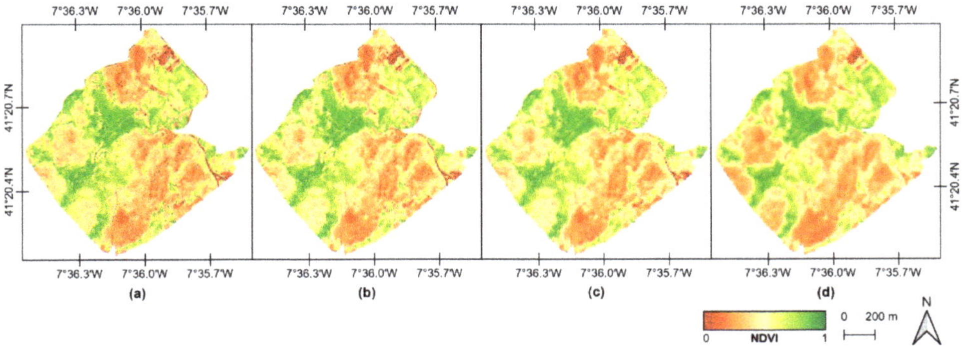

Figure 6. Normalized difference vegetation index (NDVI) computed from the multispectral data obtained from the unmanned aerial vehicle at 0.25 m (**a**) and the resamples to 5 m (**b**) and 10 m (**c**), as well as the NDVI computed from the Sentinel-2 MSI data (**d**).

Table 2. Basic statistics of the normalized difference vegetation index of the different UAV-based spatial resolutions and the Sentinel-2.

	Num. of Pixels	Minimum	Mean	Maximum	STD
UAV 0.25 m	1690×10^4	-0.39	0.51	0.99	0.23
UAV 5 m	4.57×10^4	-0.10	0.51	0.93	0.21
UAV 10 m	1.14×10^4	-0.09	0.51	0.91	0.20
Sentinel-2	1.14×10^4	-0.06	0.49	0.92	0.20

The confusion matrix presented in Table 3 shows the correlation between all NDVIs. All resolutions of the UAV-based NDVIs showed a good correlation and increased as the spatial resolution became closer to the satellite resolution.

Table 3. Correlation matrix between the normalized difference vegetation index of the different UAV-based spatial resolutions and the Sentinel-2.

	UAV 0.25 m	UAV 5 m	UAV 10 m	Sentinel-2A
UAV 0.25 m	1.00	-	-	-
UAV 5 m	0.85	1.00	-	-
UAV 10 m	0.91	0.93	1.00	-
Sentinel-2	0.84	0.90	0.93	1.00

Geospatial correlation was conducted using a 50 × 50 m grid, resulting in a total of 479 cells. The mean value of the satellite NDVI was compared with each UAV resolution, and the results are presented in Figure 7. The MI value for all approaches was 0.634. Generally, all approaches presented a similar behavior in the BILISA relationships, where 59% of the cells presented a p-value lower than 0.05: 81% of cells presented an HH or LL correlation (39.3% and 41.4%, respectively), 11% presented an LH correlation and only 8% presented an HL correlation.

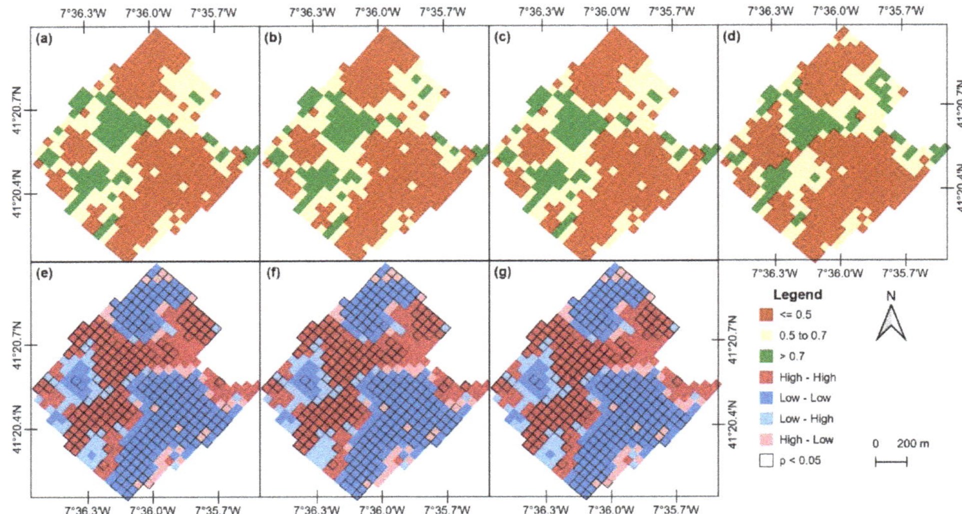

Figure 7. Mean value of the normalized difference vegetation index (NDVI) computed using the multispectral data obtained from an unmanned aerial vehicle (UAV) at 0.25 m (**a**) and the resamples to 5 m (**b**) and 10 m (**c**), as well as the NDVI computed from the Sentinel-2 Multispectral Instrument (MSI) data (**d**) in the 50 × 50 m grid. Bivariate local indicators of spatial association (BILISA) cluster maps between the NDVI map from Sentinel-2 and the UAV-based NDVIs at different spatial resolutions of 0.25 m (**e**), 5 m (**f**), and 10 m (**g**). Associations with a *p*-value < 0.05 are highlighted with a black border.

4. Discussion

This study evaluates the usage of free-access multi-temporal Sentinel-2 data to perform post-fire monitoring over an area of 361 ha in the north-eastern Portugal. The dNBI (Figure 2) was used to assess fire severity, which enabled estimation and delineation of the area affected per severity level. Both high and moderate severity classes represented the majority of the burned area (a total of 86%, corresponding to 311 ha), demonstrating a high incidence of fire disturbance in the forest stands present in the area. Moreover, both classes also presented the lowest post-fire NDVI values (Figure 4, September 2017). The same trend has been verified by other studies, noting that values decrease as fire severity rises [32]. On the other hand, unburned and low-severity areas were mostly located on the perimeter of the fire disturbance. These areas had easier access, due to the existence of roads and of priority protection by the authorities due to the proximity to settlements and infrastructures. These results are corroborated by the mean NDVI value of the multitemporal analysis (Figure 4), which shows similar values to the pre-fire data in the low-severity and unburned areas along with lower NDVI differences after the fire event (Figure 5). An example of a riparian stand that resisted fire disturbance is shown in Figure 8. On the other hand, areas classified with a high or moderate fire severity presented higher difference in the NDVI values during the analyzed period. This can be justified by the resprouting of some species and by the regeneration of others, as is the case with maritime pine, which has physical characteristics that allow its survival (thick bark and reproduction procedures) [54]. Moreover, the trend of NDVI values declining over the months can be justified by the presence of some undergrowth cover that dries out due to the absence of precipitation and increase of air temperature [55].

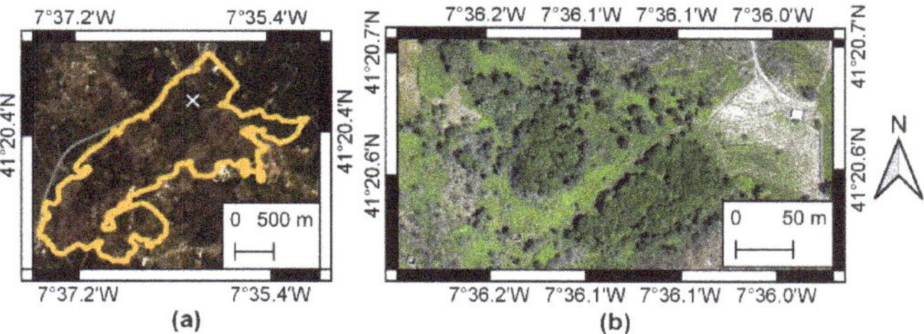

Figure 8. Overview of a stand that resisted the fire disturbance. Its location in the study area is (**a**) marked with a white cross and (**b**) visualized in the UAV-based orthophoto mosaic. Data from July 2019.

The UAV-based multispectral imagery acquired in the 116 ha of the study showed similar results when compared to the Sentinel-2 data. These findings have already been verified for WordView-2 1-m spatial resolution data [14], but never for Sentinel-2. In fact, for this type of application, the Sentinel-2 proved to be a more cost-effective approach that was able to cover wider areas, providing a short revisit time (five days) and delivering a wider spectral range. UAV-based multispectral data acquisition, on the other hand, can provide similar or higher temporal resolutions, but in a more time-consuming and expensive way, with costs increasing significantly for bigger areas [56]. This is an issue, since at least two human resources are needed who will make multiple trips and spend several days of work in order to meet a similar revisit time [57]. Furthermore, multiple batteries are needed to cover a considerable area. Fernández-Guisuraga et al. [14] used the Parrot SEQUOIA for UAV-based data acquisition during the post-fire monitoring of a 3000-ha area and faced several issues in the process. The overall procedure was time-consuming and computationally demanding, with data acquisition taking two months to conduct (resulting in a total of 100 h), and further data processing taking approximately 320 h. Some of these data then had to be discarded due to sensor malfunctions during the flights, in addition to radiometric anomalies found in the acquired images and further data storage problems. The experiment carried out by Fernández-Guisuraga et al. [14] allowed the suitability of UAV-based multispectral imagery to be determined when more information in terms of spatial variability in heterogeneous burned areas is needed. Other authors have explored fire severity measuring using UAV-based RGB imagery [17], but some limitations that directly impact its accuracy have been found such as the influence of canopy shadows, photogrammetric errors in canopy modelling and inconsistent illumination across the imagery. However, all remaining applications in terms of fire monitoring can be accomplished using satellite imagery, including those provided by Sentinel-2 MSI. Despite the great effectiveness of the satellite data for post-fire monitoring at a local/regional scale, some applications may require a significantly higher spatial resolution, making UAVs necessary, as is the case in individual tree monitoring [58], which cannot be conducted with satellite data with a decameter resolution or in real-time fire monitoring applications [12]. Thus, the complementarity of the two types of data are proven.

5. Conclusions

In this article, the potential of the use of satellite optical time series images from the ESA Copernicus Programme was addressed for monitoring relatively small areas affected by forest fires. In areas with sizes up to the one presented in this study (~400 ha), the use of small and very flexible UAVs for the analysis of post-fire vegetation recovery would be perfectly possible. However, the use of UAVs would result in a more laborious and expensive UAV tasks, requiring several visits to a field. Thus, in this study, Sentinel-2 MSI data were used to compute NBRs before and after fire disturbances in order to measure their extents and severity using difference NBR (dNBR). Subsequently, NDVI was also calculated to assess forestry recovery in the study region from 2017 to 2019. The NDVI from the Sentinel-2 MSI data was compared with UAV-based high-resolution data at different spatial resolutions

(0.25, 5 and 10 m) to access their similarities. The results demonstrated the effectiveness of satellite data for post-fire monitoring, even at a local scale. The Sentinel-2 MSI data presented a similar behavior to the UAV-based data in assessing burned areas. The confusion matrix, calculated for Sentinel-2 and UAV, showed high correlations between all NDVIs (i.e., 0.83, 0.90 and 0.93 for 0.25, 5 and 10 m spatial resolutions, respectively). Furthermore, the median and extreme values were very similar, differing no more than 0.02 for the mean, 0.04 for the minimum and 0.01 for the maximum. Thus, the availability of multi-temporal Sentinel-2 MSI data with frequent revisit times enables the severity of fire disturbances to be identified and, in a post-fire context, for the recovery of forests to be monitored and their evolutions observed when compared to pre-fire vegetation status. In this way, Sentinel-2 data can be automatically used to monitor burned areas. However, this approach should be evaluated in other areas with different fire extensions and vegetative covers, as well as in broader post-fire periods.

Author Contributions: Conceptualization, Luís Pádua and Joaquim J. Sousa; data curation, Luís Pádua; formal analysis, Luís Pádua; funding acquisition, António Sousa, Emanuel Peres and Joaquim J. Sousa; investigation, Luís Pádua, Nathalie Guimarães and Telmo Adão; methodology, Luís Pádua and Joaquim J. Sousa; project administration, Luís Pádua, Emanuel Peres and Joaquim J. Sousa; resources, Luís Pádua, Nathalie Guimarães, Telmo Adão and Joaquim J. Sousa; software, Luís Pádua, Nathalie Guimarães and Telmo Adão; supervision, António Sousa, Emanuel Peres and Joaquim J. Sousa; validation, Luís Pádua and Joaquim J. Sousa; visualization, Luís Pádua; writing—original draft, Luís Pádua; writing—review and editing, Nathalie Guimarães, Telmo Adão, António Sousa, Emanuel Peres and Joaquim J. Sousa. All authors have read and agreed to the published version of the manuscript.

Funding: Financial support provided by the FCT-Portuguese Foundation for Science and Technology (SFRH/BD/139702/2018) to Luís Pádua.

Conflicts of Interest: The authors declare no conflict of interest.

References

1. Fernandes, P.M.; Barros, A.M.G.; Pinto, A.; Santos, J.A. Characteristics and controls of extremely large wildfires in the western Mediterranean Basin. *J. Geophys. Res. Biogeosci.* **2016**, *121*, 2141–2157. [CrossRef]
2. Rego, F.C. Land Use Changes and Wildfires. In *Responses of Forest Ecosystems to Environmental Changes*; Teller, A., Mathy, P., Jeffers, J.N.R., Eds.; Springer: Dordrecht, the Netherlands, 1992; pp. 367–373. ISBN 978-94-011-2866-7.
3. Fernandes, P.M. On the socioeconomic drivers of municipal-level fire incidence in Portugal. *For. Policy Econ.* **2016**, *62*, 187–188. [CrossRef]
4. Foster, D.R.; Knight, D.H.; Franklin, J.F. Landscape Patterns and Legacies Resulting from Large, Infrequent Forest Disturbances. *Ecosystems* **1998**, *1*, 497–510. [CrossRef]
5. Mateus, P.; Fernandes, P.M. Forest Fires in Portugal: Dynamics, Causes and Policies. In *Forest Context and Policies in Portugal: Present and Future Challenges*; Reboredo, F., Ed.; World Forests; Springer International Publishing: Cham, Switzerland, 2014; pp. 97–115. ISBN 978-3-319-08455-8.
6. Departamento de Gestão de Áreas Públicas e de Proteção Florestal 10. *Relatório Provisório De Incêndios Florestais-2017*, 2017.
7. Lentile, L.B.; Holden, Z.A.; Smith, A.M.S.; Falkowski, M.J.; Hudak, A.T.; Morgan, P.; Lewis, S.A.; Gessler, P.E.; Benson, N.C. Remote sensing techniques to assess active fire characteristics and post-fire effects. *Int. J. Wildland Fire* **2006**, *15*, 319–345. [CrossRef]
8. Pádua, L.; Vanko, J.; Hruška, J.; Adão, T.; Sousa, J.J.; Peres, E.; Morais, R. UAS, sensors, and data processing in agroforestry: a review towards practical applications. *Int. J. Remote Sens.* **2017**, *38*, 2349–2391. [CrossRef]
9. Fernández-Álvarez, M.; Armesto, J.; Picos, J. LiDAR-Based Wildfire Prevention in WUI: The Automatic Detection, Measurement and Evaluation of Forest Fuels. *Forests* **2019**, *10*, 148. [CrossRef]
10. Shin, P.; Sankey, T.; Moore, M.M.; Thode, A.E. Evaluating Unmanned Aerial Vehicle Images for Estimating Forest Canopy Fuels in a Ponderosa Pine Stand. *Remote Sens.* **2018**, *10*, 1266. [CrossRef]
11. Martínez-de Dios, J.R.; Merino, L.; Caballero, F.; Ollero, A. Automatic forest-fire measuring using ground stations and Unmanned Aerial Systems. *Sensors* **2011**, *11*, 6328–6353. [CrossRef]
12. Merino, L.; Caballero, F.; Martínez-de-Dios, J.R.; Maza, I.; Ollero, A. An Unmanned Aircraft System for Automatic Forest Fire Monitoring and Measurement. *J. Intell. Robot Syst.* **2012**, *65*, 533–548. [CrossRef]

13. Yuan, C.; Zhang, Y.; Liu, Z. A survey on technologies for automatic forest fire monitoring, detection, and fighting using unmanned aerial vehicles and remote sensing techniques. *Can. J. For. Res.* **2015**, *45*, 783–792. [CrossRef]

14. Fernández-Guisuraga, J.M.; Sanz-Ablanedo, E.; Suárez-Seoane, S.; Calvo, L. Using Unmanned Aerial Vehicles in Postfire Vegetation Survey Campaigns through Large and Heterogeneous Areas: Opportunities and Challenges. *Sensors* **2018**, *18*, 586. [CrossRef] [PubMed]

15. Fraser, R.H.; Van der Sluijs, J.; Hall, R.J. Calibrating Satellite-Based Indices of Burn Severity from UAV-Derived Metrics of a Burned Boreal Forest in NWT, Canada. *Remote Sens.* **2017**, *9*, 279. [CrossRef]

16. Larrinaga, A.R.; Brotons, L. Greenness Indices from a Low-Cost UAV Imagery as Tools for Monitoring Post-Fire Forest Recovery. *Drones* **2019**, *3*, 6. [CrossRef]

17. McKenna, P.; Erskine, P.D.; Lechner, A.M.; Phinn, S. Measuring fire severity using UAV imagery in semi-arid central Queensland, Australia. *Int. J. Remote Sens.* **2017**, *38*, 4244–4264. [CrossRef]

18. Carvajal-Ramírez, F.; Marques da Silva, J.R.; Agüera-Vega, F.; Martínez-Carricondo, P.; Serrano, J.; Moral, F.J. Evaluation of Fire Severity Indices Based on Pre- and Post-Fire Multispectral Imagery Sensed from UAV. *Remote Sens.* **2019**, *11*, 993. [CrossRef]

19. Aicardi, I.; Garbarino, M.; Lingua, A.; Lingua, E.; Marzano, R.; Piras, M. Monitoring Post-Fire Forest Recovery Using Multitemporal Digital Surface Models Generated from Different Platforms. *Earsel Eproceedings* **2016**, *15*, 1–8.

20. White, R.A.; Bomber, M.; Hupy, J.P.; Shortridge, A. UAS-GEOBIA Approach to Sapling Identification in Jack Pine Barrens after Fire. *Drones* **2018**, *2*, 40. [CrossRef]

21. Hardin, P.J.; Jensen, R.R. Small-Scale Unmanned Aerial Vehicles in Environmental Remote Sensing: Challenges and Opportunities. *GIScience Remote Sens.* **2011**, *48*, 99–111. [CrossRef]

22. Chu, T.; Guo, X. Remote Sensing Techniques in Monitoring Post-Fire Effects and Patterns of Forest Recovery in Boreal Forest Regions: A Review. *Remote Sens.* **2014**, *6*, 470–520. [CrossRef]

23. Clemente, R.H.; Cerrillo, R.M.N.; Gitas, I.Z. Monitoring post-fire regeneration in Mediterranean ecosystems by employing multitemporal satellite imagery. *Int. J. Wildland Fire* **2009**, *18*, 648–658. [CrossRef]

24. Eidenshink, J.; Schwind, B.; Brewer, K.; Zhu, Z.-L.; Quayle, B.; Howard, S. A Project for Monitoring Trends in Burn Severity. *Fire Ecol.* **2007**, *3*, 3–21. [CrossRef]

25. Van Leeuwen, W.J.D. Monitoring the Effects of Forest Restoration Treatments on Post-Fire Vegetation Recovery with MODIS Multitemporal Data. *Sensors* **2008**, *8*, 2017–2042. [CrossRef] [PubMed]

26. Liu, Y.; Gong, W.; Hu, X.; Gong, J. Forest Type Identification with Random Forest Using Sentinel-1A, Sentinel-2A, Multi-Temporal Landsat-8 and DEM Data. *Remote Sens.* **2018**, *10*, 946. [CrossRef]

27. Gascon, F.; Cadau, E.; Colin, O.; Hoersch, B.; Isola, C.; Fernández, B.L.; Martimort, P. Copernicus Sentinel-2 Mission: Products, Algorithms and Cal/Val. In Proceedings of the Earth Observing Systems XIX International Society for Optics and Photonics, San Diego, CA, USA, 26 September 2014; Volume 9218, p. 92181E.

28. Drusch, M.; Del Bello, U.; Carlier, S.; Colin, O.; Fernandez, V.; Gascon, F.; Hoersch, B.; Isola, C.; Laberinti, P.; Martimort, P.; et al. Sentinel-2: ESA's Optical High-Resolution Mission for GMES Operational Services. *Remote Sens. Environ.* **2012**, *120*, 25–36. [CrossRef]

29. Moreira, F.; Ascoli, D.; Safford, H.; Adams, M.A.; Moreno, J.M.; Pereira, J.M.C.; Catry, F.X.; Armesto, J.; Bond, W.; González, M.E.; et al. Wildfire management in Mediterranean-type regions: paradigm change needed. *Environ. Res. Lett.* **2020**, *15*, 011001. [CrossRef]

30. Fernández-Manso, A.; Fernández-Manso, O.; Quintano, C. SENTINEL-2A red-edge spectral indices suitability for discriminating burn severity. *Int. J. Appl. Earth Obs. Geoinf.* **2016**, *50*, 170–175. [CrossRef]

31. Roteta, E.; Bastarrika, A.; Padilla, M.; Storm, T.; Chuvieco, E. Development of a Sentinel-2 burned area algorithm: Generation of a small fire database for sub-Saharan Africa. *Remote Sens. Environ.* **2019**, *222*, 1–17. [CrossRef]

32. Navarro, G.; Caballero, I.; Silva, G.; Parra, P.C.; Vázquez, Á.; Caldeira, R. Evaluation of forest fire on Madeira Island using Sentinel-2A MSI imagery. *Int. J. Appl. Earth Obs. Geoinf.* **2017**, *58*, 97–106. [CrossRef]

33. Filipponi, F. BAIS2: Burned Area Index for Sentinel-2. *Proceedings* **2018**, *2*, 364. [CrossRef]

34. Amos, C.; Petropoulos, G.P.; Ferentinos, K.P. Determining the use of Sentinel-2A MSI for wildfire burning & severity detection. *Int. J. Remote Sens.* **2019**, *40*, 905–930.

35. Quintano, C.; Fernández-Manso, A.; Fernández-Manso, O. Combination of Landsat and Sentinel-2 MSI data for initial assessing of burn severity. *Int. J. Appl. Earth Obs. Geoinf.* **2018**, *64*, 221–225. [CrossRef]

36. Mallinis, G.; Mitsopoulos, I.; Chrysafi, I. Evaluating and comparing Sentinel 2A and Landsat-8 Operational Land Imager (OLI) spectral indices for estimating fire severity in a Mediterranean pine ecosystem of Greece. *GIScience Remote Sens.* **2018**, *55*, 1–18. [CrossRef]

37. Filipponi, F. Exploitation of Sentinel-2 Time Series to Map Burned Areas at the National Level: A Case Study on the 2017 Italy Wildfires. *Remote Sens.* **2019**, *11*, 622. [CrossRef]

38. Chrysafis, I.; Christopoulou, A.; Kazanis, D.; Farangitakis, G.P.; Mallinis, G.; Mitsopoulos, I.; Arianoutsou, M.; Vassilakis, E.; Antoniou, V.; Theofanous, N.; et al. Post-fire vegetation recovery mapping using multi-temporal Sentinel-2A imagery in Chios island, Greece. In Proceedings of the EGU General Assembly Conference Abstracts, Vienna, Austria, 4–13 April 2018; Volume 20, p. 7066.

39. ICNG—Instituto da Convervação da Natureza e das Florestas Mapas—ICNF. Available online: http://www2.icnf.pt/portal/florestas/dfci/inc/mapas (accessed on 9 January 2020).

40. Louis, J.; Debaecker, V.; Pflug, B.; Main-Knorn, M.; Bieniarz, J.; Mueller-Wilm, U.; Cadau, E.; Gascon, F. SENTINEL-2 SEN2COR: L2A Processor for Users. In Proceedings of the Living Planet Symposium 2016, Prague, Czech Republic, 9–13 May 2016; Ouwehand, L., Ed.; Spacebooks Online: Prague, Czech Republic, 2016; Volume SP-740, pp. 1–8.

41. Key, C.; Benson, N. Landscape Assessment: Ground measure of severity, the Composite Burn Index; and Remote sensing of severity, the Normalized Burn Ratio. In *FIREMON: Fire Effects Monitoring and Inventory System*; USDA Forest Service, Rocky Mountain Research Station: Ogden, UT, USA, 2006; p. LA 1-51.

42. Veraverbeke, S.; Lhermitte, S.; Verstraeten, W.W.; Goossens, R. A time-integrated MODIS burn severity assessment using the multi-temporal differenced normalized burn ratio (dNBRMT). *Int. J. Appl. Earth Obs. Geoinf.* **2011**, *13*, 52–58. [CrossRef]

43. French, N.H.F.; Kasischke, E.S.; Hall, R.J.; Murphy, K.A.; Verbyla, D.L.; Hoy, E.E.; Allen, J.L. Using Landsat data to assess fire and burn severity in the North American boreal forest region: an overview and summary of results. *Int. J. Wildland Fire* **2008**, *17*, 443–462. [CrossRef]

44. Rouse, J.W., Jr.; Haas, R.H.; Schell, J.A.; Deering, D.W. Monitoring Vegetation Systems in the Great Plains with Erts. *NASA Spec. Publ.* **1974**, *351*, 309.

45. Gouveia, C.; DaCamara, C.C.; Trigo, R.M. Post-fire vegetation recovery in Portugal based on spot/vegetation data. *Nat. Hazards Earth Syst. Sci.* **2010**, *10*, 673–684. [CrossRef]

46. Teodoro, A.; Amaral, A. A Statistical and Spatial Analysis of Portuguese Forest Fires in Summer 2016 Considering Landsat 8 and Sentinel 2A Data. *Environments* **2019**, *6*, 36. [CrossRef]

47. Keeley, J.E. Fire intensity, fire severity and burn severity: a brief review and suggested usage. *Int. J. Wildland Fire* **2009**, *18*, 116–126. [CrossRef]

48. GRASS Development Team. *Geographic Resources Analysis Support System (GRASS) Software*; Version 7.2.; Open Source Geospatial Foundation: Beaverton, OR, USA, 2017.

49. Conrad, O.; Bechtel, B.; Bock, M.; Dietrich, H.; Fischer, E.; Gerlitz, L.; Wehberg, J.; Wichmann, V.; Böhner, J. System for Automated Geoscientific Analyses (SAGA) v. 2.1.4. *Geosci. Model Dev.* **2015**, *8*, 1991–2007. [CrossRef]

50. Moran, P.a.P. NOTES ON CONTINUOUS STOCHASTIC PHENOMENA. *Biometrika* **1950**, *37*, 17–23. [CrossRef] [PubMed]

51. Anselin, L. Local indicators of spatial association—LISA. *Geogr. Anal.* **1995**, *27*, 93–115. [CrossRef]

52. Anselin, L.; Rey, S.J. *Modern Spatial Econometrics in Practice: A Guide to GeoDa, GeoDaSpace and PySAL*; Geoda Press LLC: Chicago, IL, USA, 2014; ISBN 978-0-9863421-0-3.

53. Anselin, L.; Syabri, I.; Kho, Y. GeoDa: An Introduction to Spatial Data Analysis. *Geogr. Anal.* **2006**, *38*, 5–22. [CrossRef]

54. Fernandes, P.M.; Rigolot, E. The fire ecology and management of maritime pine (Pinus pinaster Ait.). *For. Ecol. Manag.* **2007**, *241*, 1–13. [CrossRef]

55. Šraj, M.; Brilly, M.; Mikoš, M. Rainfall interception by two deciduous Mediterranean forests of contrasting stature in Slovenia. *Agric. For. Meteorol.* **2008**, *148*, 121–134. [CrossRef]

56. Matese, A.; Toscano, P.; Di Gennaro, S.F.; Genesio, L.; Vaccari, F.P.; Primicerio, J.; Belli, C.; Zaldei, A.; Bianconi, R.; Gioli, B. Intercomparison of UAV, Aircraft and Satellite Remote Sensing Platforms for Precision Viticulture. *Remote Sens.* **2015**, *7*, 2971–2990. [CrossRef]

57. Anderson, K.; Gaston, K.J. Lightweight unmanned aerial vehicles will revolutionize spatial ecology. *Front. Ecol. Environ.* **2013**, *11*, 138–146. [CrossRef]

58. Pádua, L.; Adão, T.; Guimarães, N.; Sousa, A.; Peres, E.; Sousa, J.J. Post-fire forestry recovery monitoring using high-resolution multispectral imagery from unmanned aerial vehicles. In Proceedings of the ISPRS—International Archives of the Photogrammetry, Remote Sensing and Spatial Information Sciences, Prague, Czech Republic, 3–6 September 2019; Copernicus GmbH: Göttingen, Germany, 2019; Volume XLII-3-W8, pp. 301–305.

International Journal of
Geo-Information

Article

Recent Sea Level Change in the Black Sea from Satellite Altimetry and Tide Gauge Observations

Nevin Betül Avşar * and Şenol Hakan Kutoğlu

Department of Geomatics Engineering, Zonguldak Bülent Ecevit University, Incivez, 67100 Zonguldak, Turkey;
shakan.kutoglu@beun.edu.tr
* Correspondence: nb_avsar@beun.edu.tr

Received: 30 January 2020; Accepted: 22 March 2020; Published: 24 March 2020

Abstract: Global mean sea level has been rising at an increasing rate, especially since the early 19th century in response to ocean thermal expansion and ice sheet melting. The possible consequences of sea level rise pose a significant threat to coastal cities, inhabitants, infrastructure, wetlands, ecosystems, and beaches. Sea level changes are not geographically uniform. This study focuses on present-day sea level changes in the Black Sea using satellite altimetry and tide gauge data. The multi-mission gridded satellite altimetry data from January 1993 to May 2017 indicated a mean rate of sea level rise of 2.5 ± 0.5 mm/year over the entire Black Sea. However, when considering the dominant cycles of the Black Sea level time series, an apparent (significant) variation was seen until 2014, and the rise in the mean sea level has been estimated at about 3.2 ± 0.6 mm/year. Coastal sea level, which was assessed using the available data from 12 tide gauge stations, has generally risen (except for the Bourgas Station). For instance, from the western coast to the southern coast of the Black Sea, in Constantza, Sevastopol, Tuapse, Batumi, Trabzon, Amasra, Sile, and Igneada, the relative rise was 3.02, 1.56, 2.92, 3.52, 2.33, 3.43, 5.03, and 6.94 mm/year, respectively, for varying periods over 1922–2014. The highest and lowest rises in the mean level of the Black Sea were in Poti (7.01 mm/year) and in Varna (1.53 mm/year), respectively. Measurements from six Global Navigation Satellite System (GNSS) stations, which are very close to the tide gauges, also suggest that there were significant vertical land movements at some tide gauge locations. This study confirmed that according to the obtained average annual phase value of sea level observations, seasonal sea level variations in the Black Sea reach their maximum annual amplitude in May–June.

Keywords: Black sea; sea level change; tide gauge; satellite altimetry; GNSS

---

1. Introduction

Sea level changes occur at various time scales. Throughout geologic eras, sea levels have changed drastically many times, primarily following tectonic processes and glacial cycles [1]. During the Last Glacial Maximum (LGM), sea levels were about 130 m lower than today, because of the large amount of water held by glaciers and ice sheets [2]. After the major deglaciation (~21,000 years ago), sea levels have remained almost stable over the last 2–3 millennia [3,4]. However, with the beginning of the industrial age (late 18th to early 19th century), global sea level rise has accelerated [5–9], triggered by abrupt changes in temperature, ice cover, precipitation, etc., rather than being part of a natural cycle. Furthermore, if considering possible greenhouse gas concentration scenarios, by the end of the 21st century, global mean sea levels may rise in the range of 43 cm to 84 cm [10].

Regional sea level changes deviate substantially from that of the global mean, and some regions even reveal a condition opposite that of the global trend [11]. In this case, in addition to the global sea level change and its causes, it is essential to understand the regional variability in rates of this change (i.e., its evolution with time and space and its drivers) in order to assess the potential impacts of sea level rise in coastal areas [12]. General forcing for regional (or local) sea level patterns can be basically linked to (1) surface warming and cooling of the ocean, (2) exchange of freshwater with the atmosphere and land through evaporation, precipitation, and runoff, and (3) changes in the surface wind stress [11]. The complex response of the ocean to these forcing mechanisms causes changes in ocean circulation (hence density) and mass transport.

Sea level rise poses a significant threat to areas with low topography such as coasts, islands, and deltas. From the past to present, coastal regions have always attracted high interest in terms of their social and economic impacts [13]. Increasing human migration to these regions has made the possible consequences of sea level rise even more important. Flooding, inundation, storm, erosion, habitat loss, ecosystem damage, and contamination of underground water are the most damaging/catastrophic effects of sea level rise in coastal areas. The importance of these effects depends on the character of the coastal environment. Nevertheless, it is clear that some of them can threaten human life and coastal installations [10,14–17]. Eventually, in the near-future, rising sea levels and potentially more intense storms will exacerbate possible consequences, and more frequent extreme sea level events will occur. Therefore, effective management and sustainable use of coastal areas need multidisciplinary studies about the reasons and effects of sea level rise.

Tide gauges are one of the oldest instruments for measuring sea level changes [8]. A tide gauge measures the sea level relative to a fixed point on land, and therefore vertical movement of the point affects sea level measurements. It is necessary to perform geodetic measurements to determine sea level changes, independent of land movements at tide gauge stations. Ideally, Global Navigation Satellite System (GNSS) equipment is attached directly to the tide gauge or located nearby [18]. In addition to this, a network consisting of tide gauge stations that are referenced to the same datum and have a good distribution is needed to monitor long-term sea level changes. With the development of satellite systems, satellite altimetric techniques have been used for sea level measurements. Since 1993, the modern satellite altimetry record has provided accurate measurements of sea surface height with near-global coverage within latitudes of about 60° N and S. This technique is based on high precision measurement of the distance between the satellite and sea surface. Sea surface height is achieved by combining this information with precise satellite positional data [19].

The Black Sea in southeastern Europe (Figure 1), which is semi-enclosed, has different characteristics from other seas. It is an isolated deep body of water (average depth ~1200 m) with a restricted saltwater exchange with the Mediterranean Sea through the Turkish Straits System (the Bosporus Strait–the Sea of Marmara–the Dardanelles Strait). Unlike the Mediterranean Sea (a concentration basin), the Black Sea is an estuarine basin fed by major European rivers [20]. Additionally, another feature supported by these conditions is that the Black Sea has a specific density stratification separated by a permanent halocline [21]. Due to its geographical location, the Black Sea has been of immense strategic importance over the centuries. Its coasts have favorable natural conditions in terms of ecosystem, warm climate, fertile soils, etc., so from antiquity to the present, it has been a desirable region for human habitation [22]. Consequently, the Black Sea and its coastal zone are very sensitive to climate change and anthropogenic forcing, and thus is an area that has attracted the considerable interest of scientists.

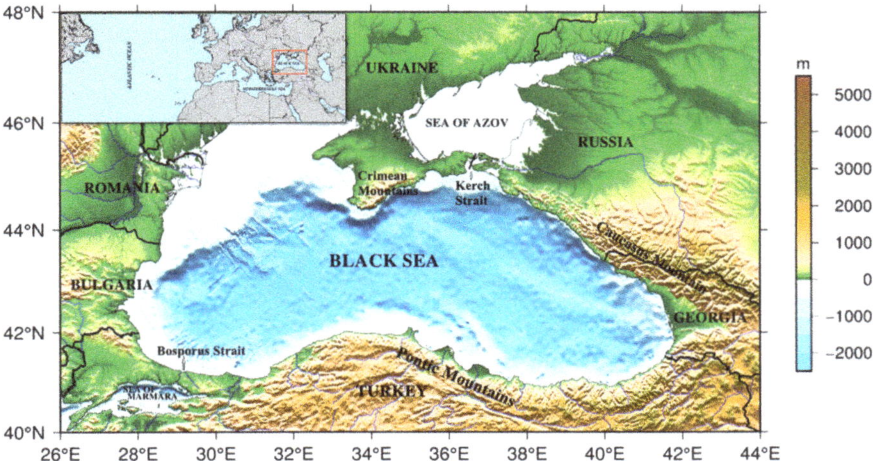

Figure 1. Black Sea with its location. Topographic data were used from the ETOPO1 1 Arc-Minute Global Relief Model bedrock data [23].

Coastal erosion and saltwater intrusion are major threats for the Black Sea coasts [24–27]. It is known that an important part of the most critical coastal erosion areas in Europe is the Black Sea coastline [28], and in particular, some coastal zones such as those in Bulgaria, Romania, and Turkey have far less protection than in other places.

Both tide gauge and altimetry observations show that sea level trends in the Black Sea vary over time. However, from the beginning of available tide gauge observations in the Black Sea, that is from the 1860s to the first decade of the 21st century, on average, an increase in sea level has been generally observed, with alternating periods of rise and fall. The Black Sea level has increased by 20 cm in the last 100 years [29]. A rise in the mean sea level of 1.83 ± 0.7 mm/year from the mid-1920s to about 1985 was mentioned in [30,31]. Forty-seven tide gauge observations, which were collected along the Black Sea coast except for the Anatolian coast before the year 1985, were evaluated by Boguslavsky et al. [32]. Considering the effects of continental discharge, atmospheric pressure, and density distribution, they asserted that the Black Sea mean level rate was 1.6 mm per year during the observational period. A rate of 2.2 mm/year from 1960 to the early 1990s was also determined by Tsimplis and Baker [33]. A rate of increase in the Black Sea level of 27.3 ± 2.5 mm/year for a period of six years (1993–1998), determined from satellite altimetry and tide gauge data, was estimated in [34]. Sea level change in the Black Sea obtained from along-track altimetry data indicated that sea level rose at a rate of 13.4 ± 0.11 mm/year over 1993–2008; in the western and eastern regions, this rate became 14.2 ± 0.16 and 12.8 ± 0.12 mm/year, respectively [30]. In the same study, the in situ and satellite results were compared and correlation coefficients ranging 0.4 to 0.7 were calculated between the tide gauge and altimeter measurements. Minimum values were obtained for tide gauges at the western and eastern coasts, whereas maximum ones were at the northern and southern coasts. The altimeter-derived Black Sea levels and corresponding independent in situ measurements have also been compared in other studies. Reasonable correlations between the data from tide gauges and the matching TOPEX/Poseidon along-track passes in the period 1992–1996 were found by Stanev et al. [35], which were 0.76, 0.68, 0.65, and 0.51 at the Tuapse, Bourgas, Varna, and Nesebar Stations, respectively. From the comparison of data over 1992–1998, the following correlation coefficients were achieved by Goryachkin and Ivanov [31], as referred to in Ginzburg et al. [30]: 0.93 for Sevastopol, 0.92 for Yalta, and 0.77 for Tuapse. High correlation coefficients varying from 0.66 to 0.89 at the tide gauge stations along the Black Sea coast (except for Batumi) have also been reported for the changing periods during 1993–2014 using gridded altimetry data [36].

Regarding coastal sea level changes, the highest change was recorded at the Poti tide gauge station (8.2 mm/year) along the Black Sea coast, whereas the lowest change was recorded at Kerci tide gauge station (1.3 mm/year) between 1860 and 1990 [37]. While at the Varna, Constantza, Sulina, Odessa, and Sevastopol Stations, the rates of sea level rise were 3.3, 2.7, 3.7, 7.1, and 3.0 mm/year, respectively; the mean subsidence rates were about 5.2, 1.1, and 6.5 mm/year at Odessa, Sevastopol and Poti, respectively. At Samsun Station, the sea level fell at a rate of −6.9 mm/year from 1963 to 1977. In order for a comparison with tide gauge records, the altimetry time series at the closest grid points to the tide gauge locations along the Black Sea coast were analyzed over the common data periods by Avsar et al. [38]. For stations with long-term records such as Poti and Tuapse, the rates of sea level changes from the satellite altimetry and tide gauges showed good agreement, and by considering the vertical land motion, the results were greatly improved. Kubryakov and Stanichnyi [39] asserted that due to the cyclonic rim current intensification for the period of 1992–2005, the sea level was rising 1.5–2 times faster in areas close to the shore than in the offshore (8–9 mm/year versus 4.5–6 mm/year). The spatial distribution of the Black Sea level trends over 1993–2014 showed that rates of sea level change during this period varied from 0.2 to 5.0 mm/year [40,41]. The southeastern region indicated a faster rise than in the other parts. Kubryakov et al. [41] pointed out that the spatial differences observed in the sea level rise were again related to basin dynamics on account of the intensification of cyclonic wind curl (3.2–4 mm/year in the coastal areas versus 1.5–2.5 mm/year in the offshore area).

In order to estimate and model regional sea level change accurately, it is important to detect sea level forcing mechanisms. According to Volkov and Landerer [42], the forcing of sea level in the Black Sea is dominated by the basin's freshwater budget (river + precipitation inputs > evaporation output) and water exchange through the Bosporus Strait as well as depth-integrated changes in seawater density. This means that changes in the water balance are the main factors for sea level variability in the Black Sea. First, it requires an investigation of long-term total sea level change in the Black Sea. This study presents an analysis of sea level changes in the Black Sea using satellite and in situ data. It aims to provide a reliable estimate of the present-day sea level rise using the data from tide gauge stations along the Black Sea coast and satellite altimetry. This study including information on absolute sea level change obtained from satellite data in the Black Sea, contributes to the relative sea level estimates by Avsar and Kutoglu [43]. Sea level observations from satellite altimetry as well as tide gauge stations have been used to infer trends in changes in Black Sea levels and their periodicity. In addition, in order to determine vertical land motion along the Black Sea coast, the data of six continuous GNSS stations, which are nearly co-located with the available tide gauge locations, were used in this study. Thus, the contribution of land motion to the coastal sea level change was also investigated.

2. Methodology and Data

2.1. Method: Harmonic Analysis

In the study, first, outlier detection was performed using the 3σ-rule. Then, monthly averaged time series of the sea level observations were obtained to provide concurrent analysis. The harmonic analysis method is preferred for sea level time series analysis, as suggested in Avsar et al. [38] and Feng et al. [44]. Harmonic analysis describes periodically recurrent phenomena and allows for the analysis of sinusoidal variation with time. Sea level changes have a periodic character. In this context, sea level time series exhibit a strong seasonality as well as a linear trend [45]. Accordingly, the seasonal variation of sea level time series can be determined by harmonic analysis, which is expressed as the sum of a number of sine and cosine terms. Simple linear regression can be used to determine the long-term trend in the sea level time series. So, a model including seasonal components (annual and semi-annual harmonics) and linear trend was used in this study [44]:

$$M(t) = M(t_0) + v(t - t_0) + \sum_{k=1}^{2} A_k \cos(\omega_k(t - t_0) + \varphi_k) + \varepsilon(t) \tag{1}$$

where $M(t)$ is the sea level time series; t is the time; t_0 is the beginning time (for example for altimetry: 1 January 1993); $M(t_0)$ is the mean sea level at t_0; v is the rate of sea level change (linear trend); $k = 1$ is the annual signal; $k = 2$ is the semi-annual signal; A is the amplitude; ω is the angular frequency; φ is the phase; and $\varepsilon(t)$ is the unmodelled residual term. Here, in order to estimate trend, phase, and amplitude, the least squares method was employed [46]. In order to determine the vertical velocities of the GNSS stations in this study, the same approaches as for the sea level changes were used.

2.2. Satellite Altimetry Observations in the Black Sea

Satellite altimetry measurements have improved our understanding of how sea levels are changing regionally [47]. The Data Unification and Altimeter Combination System (DUACS) is one of the processing systems used to produce altimetry sea level products. Here, along-track (L3) and gridded (L4) sea level products are processed over different regions and in near real time (NRT) and delayed time (DT) conditions [48]. These products were previously distributed by the French Archiving, Validation and Interpretation of Satellite Oceanographic Data (AVISO) [49], and are now produced and distributed as part of the Copernicus Marine Environment Monitoring Service (CMEMS) [50]. In this study, multi-mission (gridded) satellite altimetry data were used to investigate mean sea level change throughout the Black Sea. These grid data provide more acceptable sampling achieved by pooling measurements in a given range of latitudes and longitudes in comparison to the along-track data [51]. Additionally, the merged datasets enable high resolution sea surface height measurements [49].

The altimetry dataset in this study was daily sea surface heights from 1 January 1993 to 15 May 2017 for the Black Sea, provided by the CMEMS. These data in delayed-time are gridded (1/8° by 1/8°) sea level anomalies (SLA)s computed with respect to a twenty-year 2012 mean. SLAs have been estimated by optimal interpolation, merging the measurements from different altimeter missions: Jason-3, Sentinel-3A, HY-2A, Saral/AltiKa, Cryosat-2, Jason-2, Jason-1, TOPEX/Poseidon, ENVISAT, GFO, and ERS1/2. Necessary geophysical (solid earth, ocean and pole tides, ocean tide loading effect, sea state bias, and inverse barometer response of the ocean) and atmospheric (ionosphere, and dry/wet troposphere effects) corrections have been applied to the dataset by the data center [48]. Further information on the data can be found in Copernicus [50].

The Black Sea area comprises 3249 altimetric grid points in total. For the evaluation, at each grid point, monthly averages were computed from the daily altimetry data, and then the monthly mean sea level changes over the entire Black Sea were obtained by averaging spatially. Figure 2 shows the evolution over time of the Black Sea level based on the monthly averages from January 1993 to May 2017.

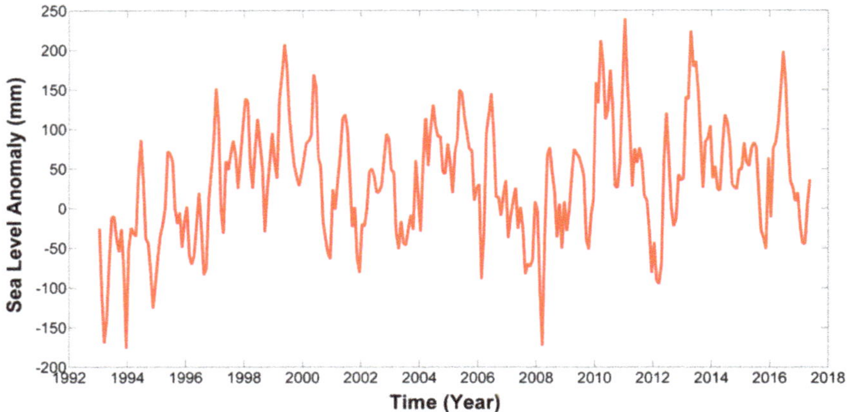

Figure 2. Monthly sea level time series from January 1993 to May 2017.

2.3. Tide Gauge Records along the Black Sea Coast

While satellite altimetry data have been available since early 1993, the availability of long-term data at many tide gauge stations is still one of the most important reasons for using these stations in sea level measurements. The Permanent Service for Mean Sea Level (PSMSL) [52] is a global data center that is responsible for the collection, publication, analysis, and interpretation of sea level data from global tide gauge station networks like the Global Sea Level Observing System (GLOSS) Core Network [53]. Sea level monitoring in Turkey is carried out by the General Directorate of Mapping (GDM) within the Turkish National Sea Level Monitoring System (TUDES) [54] stations that are in accordance with the GLOSS standards. Nevertheless, note that the amount of obtainable data from the tide gauge stations is limited. The poor spatial distribution of tide gauge stations along the coasts is a common problem for some areas including the Black Sea. Tide gauge stations have different data quality and length of records and inhomogeneous geographical distribution, and most of the data records suffer from gaps due to reasons such as equipment failure, power failure, etc., at the stations.

In this study, data from 12 tide gauge stations (having different data length) on the Black Sea coast were used. Seven tide gauge stations (Poti, Batumi, Sevastopol, Tuapse, Varna, Bourgas, and Constantza) were chosen from the PSMSL and another five (Amasra, Igneada, Trabzon, Sinop, and Sile) were from the TUDES network. Figure 3 shows the locations of all the stations in this study, and an overview of the tide gauges is given in Table 1. The Revised Local Reference (RLR) (a common datum performed by the PSMSL for each tide gauge station is approximately 7000 mm below mean sea level to avoid negative numbers in a sea level time series) data from the PSMSL are the monthly averaged time series, spanning from 65 to 140 years in the period of 1874–2013. The TUDES data are released every 15 min in the Turkish National Vertical Control Network-1999 (TUDKA-99) datum. The TUDKA-99 datum was defined based on the International Terrestrial Reference Frame-1996 (ITRF96). In the study, the monthly averaged time series of the TUDES data were derived at each station. The record with the longest time period among these stations extends to mid-2001 at Amasra. The sea level time series from some tide gauge stations have observation gaps, for example, nearly 13.4% of the records at Batumi and nearly 13.7% of those at Bourgas are void (see Table 1). Figure 4a–c show the sea level time series at the Poti, Bourgas, and Igneada tide gauges, respectively.

Vertical Coordinate Time Series of Global Navigation Satellite System (GNSS) Stations near the Tide Gauge Stations

Along the Black Sea coast, the number of GNSS receivers/stations attached directly to the tide gauge or located nearby are very sparse. Thus, the GNSS stations (Figure 3) used in this study were chosen by considering their proximity to the tide gauges (see Table 3). Data were used from six GNSS stations, which are nearly co-located (located within less than 3 km) at a tide gauge station along the Black Sea coast. The vertical displacement time series of three (TUAP, VARN, and BUR3) of these GNSS stations were provided by the Nevada Geodetic Laboratory (NGL) [55]. Another three stations (TRBN, SINP, and SLEE) are continuous GNSS stations from the Turkish National Permanent Real Time Kinematic Network (TUSAGA-Active) [56]. Their vertical coordinate time series from 2009 to 2014 were obtained (data of the related stations were processed using the GAMIT/GLOBK software in Avsar et al. [36]). Note that the TUSAGA-Active GNSS network is in ITRF96, while the positions provided from the NGL are in the International GNSS Service (IGS)14 reference frame based on the ITRF2014.

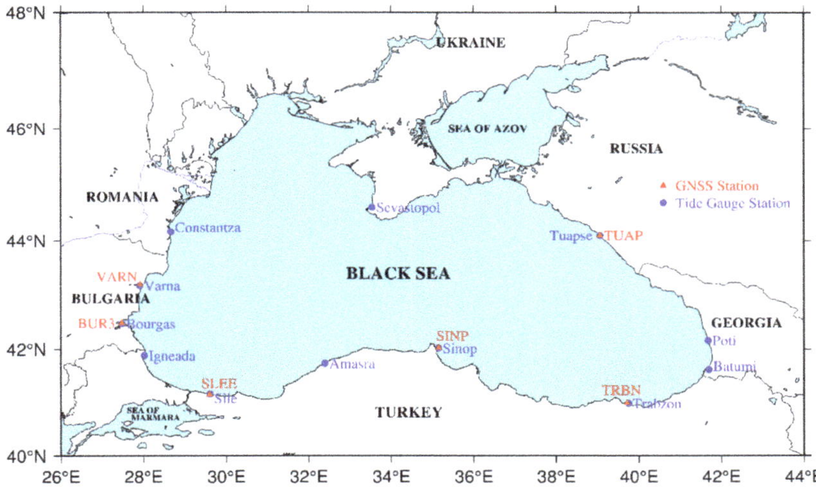

Figure 3. Locations of the tide gauges and Global Navigation Satellite System (GNSS) stations used in the study.

Table 1. General information on all the tide gauge stations used in this study.

Tide Gauge Station (Country)	Location		Data Period	Data Gaps (%)
	Latitude	Longitude		
Poti (Georgia)	42°10′ N	41°41′ E	Jan. 1874–Dec. 2013	~5.7
Batumi (Georgia)	41°38′ N	41°42′ E	Jan. 1882–Dec. 2013	~13.4
Sevastopol (Ukraine)	44°37′ N	33°32′ E	Jan. 1910–Dec. 1994	~3.1
Tuapse (Russia)	44°06′ N	39°04′ E	Jan. 1917–Dec. 2011	~1.0
Varna (Bulgaria)	43°11′ N	27°55′ E	Jan. 1929–Dec. 1996	~4.9
Bourgas (Bulgaria)	42°29′ N	27°29′ E	Jan. 1929–Dec. 1996	~13.7
Constantza (Romania)	44°10′ N	28°40′ E	Jan. 1933–Dec. 1997	~5.0
Amasra (Turkey)	41°45′ N	32°24′ E	Jun. 2001–Dec. 2014	~9.4
Igneada (Turkey)	41°53′ N	28°01′ E	Jun. 2002–Dec. 2014	~5.3
Trabzon (Turkey)	41°00′ N	39°44′ E	Jul. 2002–Dec. 2014	~0.7
Sinop (Turkey)	42°01′ N	35°09′ E	Jun. 2005–Dec. 2014	0
Sile (Turkey)	41°11′ N	29°37′ E	Jul. 2008–Dec. 2014	0

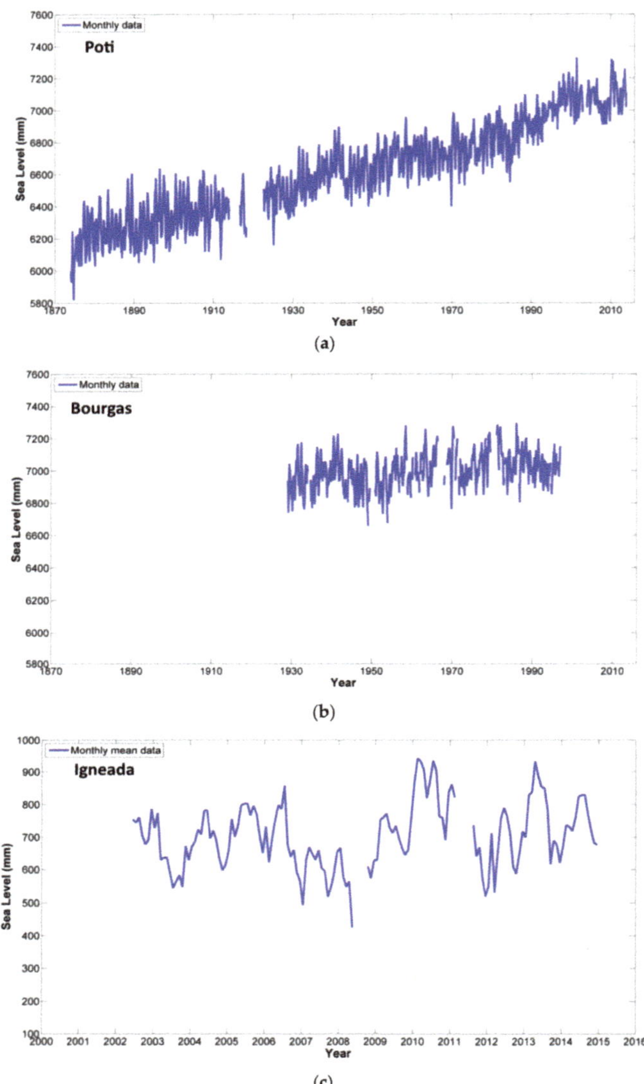

Figure 4. Three examples of the relative mean sea level changes at tide gauge stations along the Black Sea coast: (**a**) Poti from the Permanent Service for Mean Sea Level (PSMSL); (**b**) Bourgas from the PSMSL; (**c**) Igneada from the Turkish National Sea Level Monitoring System (TUDES).

3. Sea Level Changes in the Black Sea

In order to minimize the impact of low-frequency variability, records longer than 50 years should be used for long-term sea level trend estimates [8]. However, as mentioned before, although tide gauges have a good record length, they have poor spatial distribution, and only measure changes in sea level relative to the land to which they are attached. Conversely, although satellite altimetry has a short time period, it enables basin-averaged sea level change to be determined. Sea level data in the

Black Sea (especially for TUDES tide gauge stations) are mostly short-term, but nevertheless the time series indicate trends and seasonal fluctuations.

3.1. Long-Term Trend and Seasonal Variation from Satellite Altimetry Observations

The linear trend and seasonal components (annual plus semi-annual) of sea level variability in the basin average in the Black Sea were obtained through a least-squares fit of Equation (1). Here, in order to examine the long-term variability of the time series, the seasonal components were removed from the monthly values by simple subtraction of the estimates obtained by least squared fitting of seasonal sinusoids with annual and semi-annual periods (Figure 5a). The results show that the sea level in the Black Sea has risen at a rate of about 2.5 ± 0.5 mm/year between January 1993 and May 2017. In addition, the other dominant periodic behaviors (after removing seasonal cycles) in the altimetry time series were determined using Cycles Analysis & Timeseries Software (CATS) v1.0 [57]. Figure 5b demonstrates the dominant cycles over the Black Sea level time series from January 1993 to May 2017. As seen in the figure, the period of January 1993–December 2014 indicates a more apparent trend for this sea level time series. An average trend of 3.2 ± 0.6 mm/year over the Black Sea was determined for the period 1993–2014.

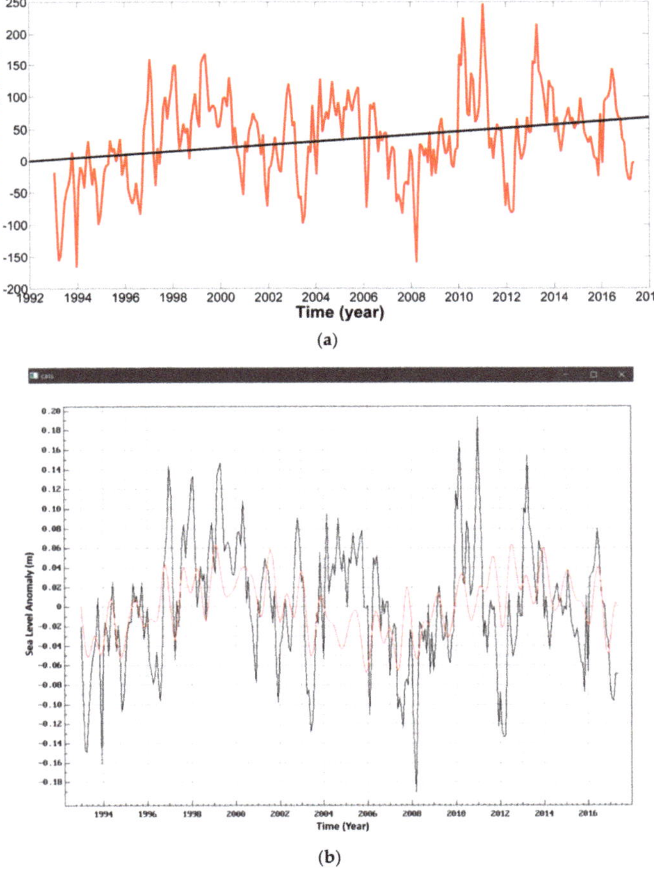

Figure 5. Monthly sea level time series in the Black Sea from 1993 to 2017 from the satellite altimetry data: (**a**) Non-seasonal sea level time series with its linear fitting. (**b**) Dominant cycles (red) over the detrended and non-seasonal sea level time series (black).

The satellite altimetry observations from 1993 to 2014 yielded a standard deviation of 7.5 cm for sea level anomalies in the Black Sea. The mean sea level anomaly in 2010 was about 20 cm above the 1993 average. This was the highest annual average in the satellite record from 1993 to the present, with the record high sea level anomalies occurring in March 2010, January 2011, and April 2013. Moreover, satellite altimetry data show there have been some strong fluctuations of sea level in the Black Sea; the difference between the mean sea level anomalies from December 2009 to January 2010 was about 15 cm.

For the seasonal components of the sea level variations in the Black Sea, the average annual and semi-annual amplitudes were detected as 38.02 ± 6.01 mm and 23.74 ± 6.01 mm, respectively, from the satellite altimetry. The average annual and semi-annual phases were 147.38 ± 0.17° (~4.9th month) and 338.04 ± 0.26° (~11.3rd month), respectively. Accordingly, the annual cycle of the sea level variations measured by the altimetry attained its maximum value in about May. Many studies have reported that the mean sea level reaches the highest levels in May–June in the Black Sea [37,58–60]. Our results confirm this condition.

3.2. Coastal Sea Level Changes from Tide Gauge Records

In the evaluation, the data gaps were excluded, and the data periods of the tide gauges were rearranged (see Table 2). However, data with less than four consecutive missing months were used through linear interpolating. Eventually, Equation (1) was used in the analysis of the sea level time series from the tide gauge stations. As an example, the trend and harmonic model of the Amasra tide gauge station are shown in Figure 6.

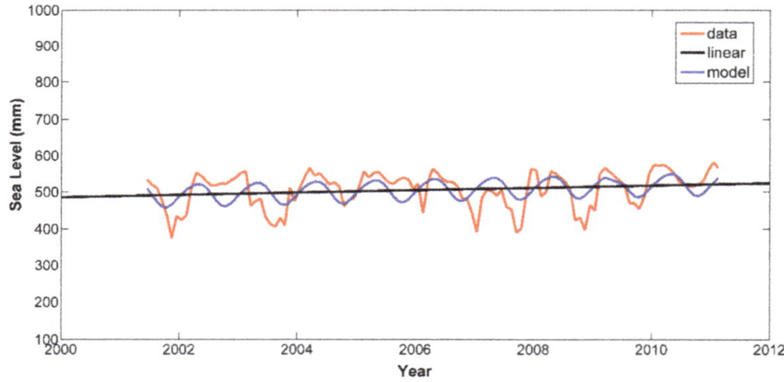

Figure 6. Trend and harmonic model of ~10-year sea level time series at the Amasra tide gauge station.

After removing seasonal variations, the trend of each tide gauge station along the Black Sea coast was derived from the varying record lengths considering the data gaps. The linear variation with time and seasonal components of the observed sea level in the tide gauge stations are given in Table 2. Accordingly, the results show that the rates of the sea level change vary from coast to coast. Consequently, nearly all the tide gauge stations (except for Bourgas) indicated rising sea levels. The greatest rise along the Black Sea coast was recorded in Poti (7.01 mm/year), and the lowest ones in Varna (1.53 mm/year) and Sevastopol (1.56 mm/year). The distribution of the data gaps in the Bourgas sea level time series (see Figure 4b) did not allow a reliable trend to estimate for this station, despite the interpolation. In addition, Sinop Station showed no significant sea level change. The non-significant results may be related to the short records, since trend estimations are sensitive to the length of the record. It would be unsafe to detect long-term trends from the short datasets [37].

As shown in Table 2, the semi-annual amplitudes of coastal sea changes were about 2–3 times smaller than the annual amplitudes. Seasonal (annual and semi-annual) sea level change signals generally reach their maximum values in May–June. However, the maximum annual amplitude of the sea level change at Amasra occurs in April, nearly one month earlier than those of the other stations.

Table 2. Trend and seasonal components of coastal (relative) sea level changes at the tide gauge stations along the Black Sea coast (the longest available data periods of the tide gauges were used for the analysis).

Tide Gauge Station	Data Period	Trend (mm/year)	Annual		Semi-Annual	
			Amplitude (mm)	Phase (°)	Amplitude (mm)	Phase (°)
Poti	Aug. 1922 Dec. 2002	7.01 ± 0.12	77.42 ± 4.05	157.76 ± 0.05	35.86 ± 4.05	26.52 ± 0.11
Batumi	Jan. 1925 Dec. 1996	3.52 ± 0.15	78.93 ± 4.43	158.48 ± 0.06	35.19 ± 4.43	22.01 ± 0.13
Sevastopol	Sep.1944 Dec. 1994	1.56 ± 0.22	79.41 ± 4.58	139.65 ± 0.06	30.07 ± 4.59	16.25 ± 0.15
Tuapse	Jan. 1943 Dec. 2011	2.92 ± 0.14	70.42 ± 3.85	142.41 ± 0.06	37.00 ± 3.85	29.78 ± 0.10
Varna	Jan. 1926 Nov. 1961	1.53 ± 0.48	69.54 ± 6.42	152.73 ± 0.09	27.38 ± 6.41	344.06 ± 0.23
Bourgas	Feb. 1981 Jan. 1996	−7.52 ± 1.33	67.23 ± 8.13	141.78 ± 0.12	20.84 ± 8.12	19.83 ± 0.39
Constantza	Jan. 1945 Dec. 1979	3.02 ± 0.46	78.14 ± 6.55	127.94 ± 0.08	15.74 ± 6.55	26.34 ± 0.42
Amasra	Jun. 2001 Feb. 2011	3.43 ± 1.42	30.69 ± 5.71	104.70 ± 0.18	3.47 ± 5.66	340.66 ± 1.63
Igneada	Jun. 2002 Dec. 2014	6.94 ± 2.18	49.16 ± 11.17	130.14 ± 0.23	16.01 ± 11.22	50.66 ± 0.70
Trabzon	Jul. 2002 Dec. 2014	2.33 ± 1.75	62.77 ± 8.93	153.45 ± 0.14	27.09 ± 8.93	17.09 ± 0.33
Sinop	Jun. 2005 Dec. 2014	0.43 ± 2.88	49.04 ± 11.26	135.82 ± 0.23	29.53 ± 11.28	12.06 ± 0.38
Sile	Jul. 2008 Dec. 2014	5.03 ± 4.84	62.92 ± 12.84	128.86 ± 0.20	22.90 ± 12.84	49.11 ± 0.56

In many studies, the results showed that tide gauge records along the Black Sea coast were in reasonable agreement with the satellite altimeter observations [30,31,35,36]. Nevertheless, in order to detect absolute sea level changes in the tide gauge locations, the vertical land motions obtained from GNSS measurements should be separated from sea level records [18,36,38]. In this study, unfortunately, the data periods of the GNSS time series did not exactly coincide with those of the tide gauges (see Table 3). However, the GNSS-derived estimates can give information on the recent land motions along the Black Sea coast. Here, the vertical coordinate time series from the nearby GNSS stations were analyzed using Equation (1), and thus the vertical velocities of these six GNSS stations were estimated. Accordingly, for the Tuapse, Varna, Bourgas, Trabzon, Sinop, and Sile tide gauge locations, the GNSS-derived vertical land motions are presented in Table 3, along with the distances between the tide gauge and related GNSS stations. The results in Table 3 show land subsidence motions at the Tuapse, Varna, Trabzon, and Sile locations. On the other hand, land uplift motions were seen at the Bourgas and Sinop locations. Especially for Sile, the high relative sea level rise may result from the land subsidence.

Table 3. Vertical land motions at the tide gauge locations along the Black Sea coast.

Tide Gauge Station	GNSS Station	Data Period		Distance (km)	Vertical Velocity (mm/year)
		Tide Gauge	GNSS		
Tuapse	TUAP	1943–2011	2015–2017	0.05	−1.7 ± 0.5
Varna	VARN	1926–1961	2005–2017	2.1	−1.1 ± 0.1
Bourgas	BUR3	1981–1996	2009–2014	1.5	4.2 ± 0.2
Trabzon	TRBN	2002–2014	2009–2014	2.8	−1.9 ± 0.3
Sinop	SINP	2005–2014	2010–2014	0.8	6.2 ± 2.5
Sile	SLEE	2008–2014	2009–2014	1.2	−3.0 ± 0.6

4. Discussion

In the Black Sea, having a limited interaction with the Atlantic Ocean, there are strong temporal mass variations due to its wide drainage area covering a large part of Europe and Asia, and sea level change is closely related to its hydrological balance. The results of this study confirmed that the Black Sea level has continued to rise over the near satellite altimetry era (1993–2017). In this context, monitoring sea level change in the Black Sea is critical for determining its long-term variability and mitigating its negative impacts.

Figure 7 focuses on the spatial distributions of low-lying areas surrounding the Black Sea. Since coastal slope is the main indicator, these areas are highly vulnerable to sea level rise. In order to estimate the vulnerability of these areas, the general characteristic of the regions should be examined in terms of soil type, land use, population, income, etc.

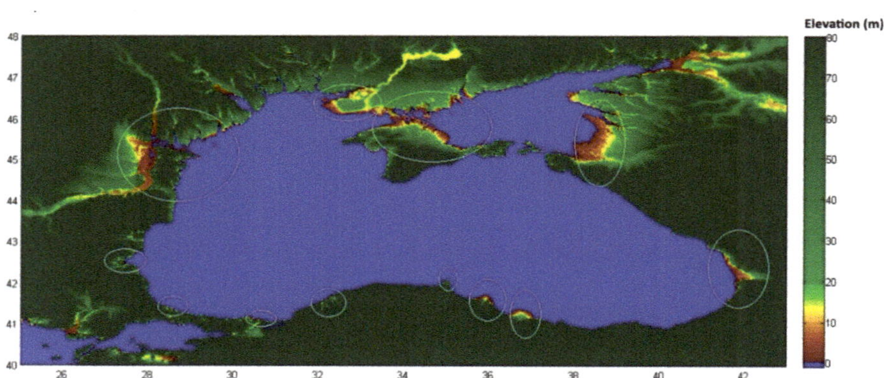

Figure 7. Areas with low slope along the Black Sea shore.

The level of the Black Sea has been rising at a mean rate of ~2.5 mm/year from January 1993 to May 2017, although a slowdown of this rate was recorded over the last about three years. Nevertheless, in order to confirm this supposition, the dominant cycles in the Black Sea level time series should be examined spectrally. Thus, the recent rate of sea level rise can be estimated more accurately. Note that, the dominant cycles of sea level change indicate that the Black Sea rose at a rate of about 3.2 ± 0.6 mm/year until December 2014. This rate was nearly identical to the global trend, which was reported by Legeais et al. [61]; this common tendency may be attributed to global warming [30]. Here, it is appropriate to summarize the available literature on the Black Sea level changes for a rightful evaluation (Table 4). When considering the rate values in Table 4 as well as the character of sea level fluctuations in the Black Sea, the sea level generally tends to rise in the long-term. The estimated rates for the short periods were higher than the estimates for the long periods (note that a period under five years is not significant for statistics). The rate of sea level rise estimated in this study for a period of approximately 22 years was about 1.8 times greater than the rate in the preceding half a century

(1920–1985), as quoted from Goryachkin and Ivanov [31]. However, the fluctuations in the Black Sea from 1993 to 2014 were not uniform: the sea level rose in general for the period of January 1993–June 1999, fell during July 1999–April 2006 (even though it slightly increased from the beginning of 2001 to about 2005), and rose again until 2014. Then, it started to fall again; throughout the 1993–2017 period mentioned in this study, the mean sea level displayed a positive trend of 2.5 ± 0.5 mm per year. According to Ginzburg et al. [30], the sea level increase from 1993 to 1999, and then decrease from 1999 to 2001 are in agreement with data on the Danube River discharge. In addition, it was mentioned in Cazenave et al. [34], Avsar et al. [62], and Vigo et al. [63] that sea level rise over 1993–1999 showed good correlation with the increase in sea surface temperature in the Black Sea over this period.

Table 4. Estimates of the rate of the basin-averaged sea level rise in the Black Sea.

Reference	Data Type	Period	Value of Rise or Linear Trend
[29]	Tide Gauge	1890–1990	20 cm
[31]	Tide Gauge	1920–1985	1.83 mm/year
[32]	Tide Gauge	1873–1985	1.6 mm/year
[33]	Tide Gauge	1960–1990	2.2 mm/year
[34]	Along-Track Altimetry	1993–1998	27.3 mm/year
[64]	In-Situ [1]	1944–2003	2.5 mm/year
[39]	Along-Track Altimetry	1992–2005	7.6 mm/year
[30]	Along-Track Altimetry	1993–2008	13.4 mm/year

[1] Steric data and precipitation/evaporation data provided from climatology and meteorological stations.

Along the coasts, complex ocean dynamics occur at shorter spatial and temporal scales. For example, tides are much more complex near the shore than in the open sea. Moreover, the high frequency variations due to atmospheric pressure and tides must be taken into account in these areas. Over shorter periods, sea levels rise faster at the coasts than offshore [65]. The main feature of sea dynamics in the Black Sea is the cyclonic rim current flowing along the continental slope, and it leads to lower sea level in the interior of the basin and higher sea level along the coast [40]. Table 5 presents an extensive review of the rates of the sea level changes at common tide gauges from different studies. Although the data periods are different, the results obtained in this study showed similar characteristics with Alpar et al. [37] and Avsar et al. [38], especially for Poti and Tuapse. At Poti Station, a high sea level trend has been generally estimated. This relative sea level rise may have resulted from subsidence at the Poti coast [30]. At Tuapse, ground subsidence may be a determinant in the sea level rise at this station [38]. While our results indicate that the rate of sea level rise during 1945–1994 slowed down at the Sevastopol tide gauge station, the trend estimates in Alpar et al. [37] and Kubryakov and Stanichnyi [39] were higher. The trend estimates in this study were calculated after removing the seasonal cycles. Thus, there was a small discrepancy in the trend estimates along the southern coast of the Black Sea compared to Avsar et al. [38].

Table 5. Estimates of the rates of sea level changes at some tide gauge stations along the Black Sea coast (the data periods from [37] are approximately known, except for Samsun).

Tide Gauge Station	Period	Reference	Linear Trend of Sea Level Change	Linear Trend of Vertical Land Motion
Poti	1890–1950	[37]	8.2 mm/year	−6.5 mm/year
	1993–2013	[38]	4.1 mm/year	
Constantza	1860–1990	[37]	2.7 mm/year	
Odessa	1870–1960	[37]	7.1 mm/year	−5.2 mm/year
	1993–2005	[39]	−4.2 mm/year	
Sevastopol	1870–1960	[37]	3.0 mm/year	−1.1 mm/year
	1993–2005	[39]	8.3 mm/year	
Tuapse	1993–2011	[38]	4.3 mm/year	

Figure 8 depicts the coastal areas in the Black Sea, which would be under water if the sea level rises 1 m. As previously outlined, coastal erosion is also a remarkable problem along the Black Sea coastline. The observed rise rates (see Table 2) along the Black Sea coast and the basin-averaged rate (from the gridded altimetry data) of the sea level rise may be significant for the threat of coastal erosion. It has been estimated by Allenbach et al. [24] that a 50 cm rise in sea level might lead to about a 50% reduction in the Black Sea beach area. According to Goryachkin and Ivanov [31], the shore might retreat 1–2 m for a rise in the sea level by 1 cm [41].

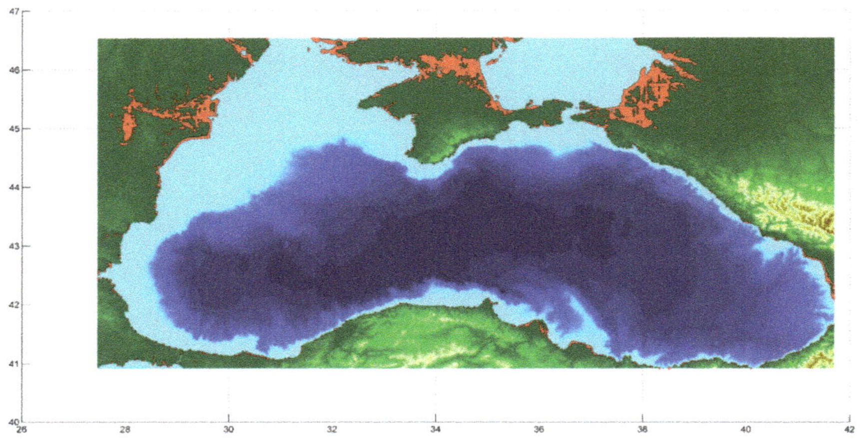

Figure 8. Coastal areas under water (marked in red), if the sea level rises 1 m.

5. Conclusions

Numerous studies have been carried out to determine the long-term variability of sea level in the Black Sea. These studies, based on altimetry and tide gauge data, have revealed that the level of the Black Sea has risen. Our study dealt with recent sea level changes in the Black Sea over a time period for which data from tide gauges and satellite altimetry are available. The mean rate of the sea level rise has been estimated as 2.5 ± 0.5 mm/year over the entire Black Sea by using the gridded satellite altimetry data covering January 1993–May 2017. During this period, it was seen that inter-annual variability of non-seasonal sea level change was quite strong (with a standard deviation of about 6.7 cm). In addition, coastal sea level changes were analyzed from 12 tide gauge stations along the Black Sea coast. However, most tide gauge data are not up to date and the spatial distribution of the stations is sparse. Nevertheless, using the available data, relative sea level changes along the Black Sea were assessed, and the results generally reflect a rise in the sea level. The highest rate of rise (7.01 ± 0.12 mm/year) was at the Poti tide gauge station. These results, combined with the vertical rates of GNSS stations, showed that at some tide gauge locations, there were significant vertical movements. This study suggests that a regional network of tide gauge stations with a suitable spatial distribution, along with co-located continuous GNSS stations along the Black Sea coastline, should be established. Continuous geodetic measurements can be used to monitor vertical land movements to estimate absolute sea level changes independent of vertical motions of the land.

The results of this study demonstrate that accurate modeling of sea level changes depending on time and location in the Black Sea, which is semi-enclosed, is crucial for risk assessments related to sea level rise, analysis of coastal change, and planning of coastal area use. Local, regional, and national patterns of potential consequences of sea level rise should be assessed, and coastal vulnerabilities should be identified in this region. The implications of sea level rise should be considered for population location, economic, infrastructure, and construction planning. This issue should be regarded as higher priority in coastal management. The related governments and local authorities should design long-term policy for coastal planning. The necessary precautions for reducing the effects of sea level rise should be implemented for all coastal areas.

Author Contributions: Conceptualization, methodology, and validation, Nevin Betül Avşar and Şenol Hakan Kutoğlu; investigation, formal analysis, writing—original draft preparation, visualization, resources, and writing—review & editing, Nevin Betül Avşar; supervision, and project administration, Şenol Hakan Kutoğlu; Both authors have read and agreed to the published version of the manuscript.

Funding: This research received no external funding.

Acknowledgments: The authors express their gratefulness to the anonymous reviewers for their valuable suggestions. The authors are also grateful to the organizations that provided the data including the Copernicus Marine Environment Monitoring Service (CMEMS), the General Directorate of Mapping (Turkey) – GDM, the Permanent Service for Mean Sea Level – PSMSL, and the Nevada Geodetic Laboratory – (NGL). The authors would like to thank Dr. Gokhan Gurbuz for his contribution in the evaluation of the GNSS data. This study was based on the first author's PhD thesis titled as "Sea level changes in the Black Sea and its impacts on the coastal areas".

References

1. Miller, K.; Mountain, G.S.; Wright, J.D.; Browning, J.V. A 180 million year record of sea level and ice volume variations from continental margin deep sea isotopic records. *Oceanography* **2011**, *24*, 40–53. [CrossRef]
2. Clark, P.U.; Mix, A.C. Ice sheets and sea level of the last glacial maximum. *Quat. Sci. Rev.* **2002**, *21*, 1–7. [CrossRef]
3. Cazenave, A. Sea Level Changes Recent Past, Present, Future. In Proceedings of the EGU General Assembly 2014 GIFT Workshop, Vienna, Austria, 27–30 April 2014. Available online: https://cdn.egu.eu/media/filer_public/2014/05/09/5_-_cazenave_anny.pdf (accessed on 27 February 2020).
4. Lambeck, K.; Chappell, J. Sea level change through the last glacial cycle. *Science* **2001**, *292*, 679–686. [CrossRef] [PubMed]
5. Church, J.A.; Clark, P.U.; Cazenave, A.; Gregory, J.M.; Jevrejeva, S.; Levermann, A.; Merrifield, M.A.; Milne, G.A.; Nerem, R.S.; Nunn, P.D.; et al. Sea level change. In *Climate Change 2013: The Physical Science Basis. Contribution of Working Group I to the Fifth Assessment Report of the Intergovernmental Panel on Climate Change*; Stocker, T.F., Qin, D., Plattner, G.-K., Tignor, M., Allen, S.K., Boschung, J., Nauels, A., Xia, Y., Bex, V., Midgley, P.M., Eds.; Cambridge University Press: Cambridge, UK; New York, NY, USA, 2013; pp. 1137–1216.
6. Church, J.A.; White, N.J. A 20th century acceleration in global sea-level rise. *Geophys. Res. Lett.* **2006**, *33*, L01602. [CrossRef]
7. Church, J.A.; White, N.J. Sea-level rise from the late 19th to the early 21st century. *Surv. Geophys.* **2011**, *32*, 585–602. [CrossRef]
8. Douglas, B.C. Sea level change in the era of the recording tide gauge. In *Sea Level Rise, History and Consequences*, 1st ed.; Douglas, B.C., Kearney, M.S., Leatherman, S.P., Eds.; Academic Press: Cambridge, MA, USA, 2001; Volume 75, pp. 37–64.
9. Nerem, R.S.; Leuliette, E.; Cazenave, A. Present-day sea-level change: A review. *CR Geosci.* **2006**, *338*, 1077–1083. [CrossRef]
10. Oppenheimer, M.; Glavovic, B.C.; Hinkel, J.; van de Wal, R.; Magnan, A.K.; Abd-Elgawad, A.; Cai, R.; CifuentesJara, M.; DeConto, R.M.; Ghosh, T.; et al. Sea level rise and implications for low-lying islands, coasts and communities. In *IPCC Special Report on the Ocean and Cryosphere in a Changing Climate*; Pörtner, H.-O., Roberts, D.C., Masson-Delmotte, V., Zhai, P., Tignor, M., Poloczanska, E., Mintenbeck, K., Alegria, A., Nicolai, M., Okem, A., et al., Eds.; IPCC: Geneva, Switzerland, 2019; (in press). Available online: https://www.ipcc.ch/site/assets/uploads/sites/3/2019/12/SROCC_FullReport_FINAL.pdf (accessed on 29 February 2020).
11. Stammer, D.; Cazenave, A.; Ponte, R.M.; Tamisiea, M.E. Causes for contemporary regional sea level changes. *Annu. Rev. Mar. Sci.* **2013**, *5*, 21–46. [CrossRef]
12. Meyssignac, B.; Cazenave, A. Sea level: A review of present-day and recent-past changes and variability. *J. Geodyn.* **2012**, *58*, 96–109. [CrossRef]
13. Creel, L. Ripple Effects: Population and Coastal Regions. Available online: https://www.prb.org/rippleeffectspopulationandcoastalregions/ (accessed on 27 February 2020).

14. Douglas, B.C.; Kearney, M.S.; Leatherman, S.P. (Eds.) *Sea Level Rise History and Consequences*, 1st ed.; Academic Press: Cambridge, MA, USA, 2001; Volume 75, p. 232.

15. Karaca, M.; Nicholls, R.J. Potential implications of accelerated sea-level rise for Turkey. *J. Coast. Res.* **2008**, *24*, 288–298. [CrossRef]

16. Nicholls, R.J. Impacts of and responses to sea level rise. In *Understanding Sea-Level Rise and Variability*; Church, J.A., Woodworth, P., Aarup, T., Wilson, W.S., Eds.; Wiley-Blackwell Publishing: London, UK, 2010; pp. 17–51.

17. Nicholls, R.J.; Cazenave, A. Sea-level rise and its impact on coastal zones. *Science* **2010**, *328*, 1517–1520. [CrossRef]

18. Woodworth, P.L.; Wöppelmann, G.; Marcos, M.; Gravelle, M.; Bingley, R.M. Why we must tie satellite positioning to tide gauge data. *Eos* **2017**, *98*, 13–15. [CrossRef]

19. Fu, L.-L.; Cazenave, A. *Satellite Altimetry and Earth Sciences: A Handbook of Techniques and Applications. International Geophysics Series*; Fu, L.-L., Cazenave, A., Eds.; Academic Press: San Diego, CA, USA, 2001; Volume 69, p. 463.

20. Jaoshvili, S. *The Rivers of the Black Sea*; European Environmental Agency Technical Report No: 71; Khomerki, I., Gigineishvili, G., Kordzadze, A., Eds.; European Environmental Agency: Copenhagen, Denmark, 2002; p. 58. Available online: https://www.eea.europa.eu/publications/technical_report_2002_71 (accessed on 5 February 2019).

21. Oguz, T.; Tugrul, S.; Kideys, A.E.; Ediger, V.; Kubilay, N. Physical and biogeochemical characteristics of the Black Sea. In *The Sea*; Robinson, A.R., Brink, K.H., Eds.; Harward University Press Cambridge: Cambridge, MA, 2006; Volume 14, pp. 1333–1371.

22. Grinevetsky, S.R.; Zonn, I.S.; Zhiltsov, S.S.; Kosarev, A.N.; Kostianoy, A.G. *The Black Sea Encyclopedia*; Springer Verlag: Berlin/Heidelberg, Germany, 2015; p. 889.

23. Amante, C.; Eakins, B.W. *ETOPO1 Arc-Minute Global Relief Model: Procedures, Data Sources and Analysis*; NOAA Technical Memorandum NESDIS NGDC-24; NOAA: Boulder, Colorado, 2009; p. 19.

24. Allenbach, K.; Garonna, I.; Herold, C.; Monioudi, I.; Giuliani, G.; Lehmann, A.; Velegrakis, A.F. Black Sea beaches vulnerability to sea level rise. *Environ. Sci. Policy* **2015**, *46*, 95–109. [CrossRef]

25. Kale, M.M.; Ataol, M.; Tekkanat, İ.S. Assessment of shoreline alterations using a digital shoreline analysis system: A case study of changes in the Yeşilırmak Delta in northern Turkey from 1953 to 2017. *Environ. Monit. Assess.* **2019**, *191*, 398. [CrossRef] [PubMed]

26. Simav, Ö. Investigation of the Potential Impacts of the Sea Level Rise on Coastal Areas. Ph.D. Thesis, Istanbul Technical University, Istanbul, Turkey, 2012. (in Turkish).

27. Tatui, F.; Pirvan, M.; Popa, M.; Aydogan, B.; Ayat, B.; Görmüș, T.; Korzinin, D.; Vaidianu, N.; Vespremeanu-Stroe, A.; Zainescu, F.; et al. The Black Sea coastline erosion: Index-based sensitivity assessment and management-related issues. *Ocean Coast. Manag.* **2019**, *182*, 104949. [CrossRef]

28. Brown, S.; Nicholls, R.J.; Vafeidis, A.; Hinkel, J.; Watkiss, P. *The Impacts and Economic Costs of Sea-Level Rise on Coastal Zones in the EU and the Costs and Benefits of Adaptation, Summary of Results from the EC RTD Climate Cost Project*; The Climate Cost Project, Final Report; Volume 1: Europe; Watkiss, P., Ed.; Stockholm Environment Institute: Stockholm, Sweden, 2011; p. 44.

29. Terziev, F.S. (Ed.) *Hydrometeorology and Hydrochemistry of the USSR Seas*; Project: Seas of the USSR; Volume IV: The Black Sea; Issue 1: Hydrometeorological Conditions; Gidrometeoizdat: St. Petersburg, Russian, 1991; p. 429. (in Russian)

30. Ginzburg, A.I.; Kostianoy, A.G.; Sheremet, N.A.; Lebedev, S.A. Satellite altimetry applications in the Black Sea. In *Coastal Altimetry*; Vignudelli, S., Kostianoy., A.G., Cipollini, P., Benveniste, J., Eds.; Springer: Berlin/Heidelberg, Germany, 2011; pp. 367–387.

31. Goryachkin, Y.N.; Ivanov, V.A. *Black Sea Level: The Past, Present and Future*; MHI NANU: Sevastopol, Russian, 2006; p. 210. (in Russian)

32. Boguslavsky, S.G.; Kubryakov, A.I.; Ivashchenko, I.K. Variations of the Black Sea level. *Phys. Oceanogr.* **1998**, *9*, 199–208. [CrossRef]

33. Tsimplis, M.N.; Baker, T.F. Sea level drop in the Mediterranean Sea: An indicator of deep water salinity and temperature changes? *Geophys. Res. Lett.* **2000**, *27*, 1731–1734. [CrossRef]

34. Cazenave, A.; Bonnefond, P.; Mercier, F.; Dominh, K.; Toumazou, V. Sea level variations in the Mediterranean Sea and Black Sea from satellite altimetry and tide gauge. *Glob. Planet. Chang.* **2002**, *34*, 59–86. [CrossRef]

35. Stanev, E.V.; Le Traon, P.-Y.; Peneva, E.L. Sea level variations and their dependency on meteorological and hydrological forcing: Analysis of altimeter and surface data for the Black Sea. *J. Geophys. Res. Ocean.* **2000**, *105*, 17203–17216. [CrossRef]

36. Avsar, N.B.; Jin, S.; Kutoglu, S.H.; Gurbuz, G. Vertical land motion along the Black Sea coast from satellite altimetry, tide gauges and GPS. *Adv. Space Res.* **2017**, *60*, 2871–2881. [CrossRef]

37. Alpar, B.; Dogan, E.; Yuce, H.; Altıok, H. Sea level changes along the Turkish coasts of the Black Sea, the Aegean Sea and the Eastern Mediterranean. *Mediterr. Mar. Sci.* **2000**, *1*, 141–156. [CrossRef]

38. Avsar, N.B.; Jin, S.; Kutoglu, H.; Gurbuz, G. Sea level change along the Black Sea coast from satellite altimetry, tide gauge and GPS observations. *Geod. Geodyn.* **2016**, *7*, 50–55. [CrossRef]

39. Kubryakov, A.A.; Stanichnyi, S.V. The Black Sea level trends from tide gauges and satellite altimetry. *Russ. Meteorol. Hydrol.* **2013**, *38*, 329–333. [CrossRef]

40. Avsar, N.B.; Kutoglu, S.H.; Erol, B.; Jin, S. Sea level changes in the Black Sea using satellite altimetry and tide gauge observations. In Proceedings of the 26th IUGG General Assembly, Prague, Czech Republic, 22 June–2 July 2015. Available online: http://www.czech-in.org/cmdownload/IUGG2015/presentations/IUGG-5542.pdf (accessed on 27 February 2020).

41. Kubryakov, A.A.; Stanichny, S.V.; Volkov, D.L. Quantifying the impact of basin dynamics on the regional sea level rise in the Black Sea. *Ocean Sci.* **2017**, *13*, 443–452. [CrossRef]

42. Volkov, D.L.; Landerer, F.W. Internal and external forcing of sea level variability in the Black Sea. *Clim. Dyn.* **2015**, *45*, 2633–2646. [CrossRef]

43. Avsar, N.B.; Kutoglu, S.H. Relative sea level change along the black sea coast from tide gauge observations. In the international archives of the photogrammetry, remote sensing and spatial information sciences. In Proceedings of the GeoInformation for Disaster Management (Gi4DM 2019), Prague, Czech Republic, 3–6 September 2019; Volume XLII-3/W8, pp. 43–47. [CrossRef]

44. Feng, G.; Jin, S.; Zhang, T. Coastal sea level changes in Europe from GPS, tide gauge, satellite altimetry and GRACE, 1993–2011. *Adv. Space Res.* **2013**, *51*, 1019–1028. [CrossRef]

45. Cazenave, A.; Cabanes, C.; Dominh, K.; Mangiarotti, S. Recent sea level changes in the Mediterranean Sea revealed by TOPEX/POSEIDON satellite altimetry. *Geophys. Res. Lett.* **2001**, *28*, 1607e10. [CrossRef]

46. Pugh, D.T. *Tides, Surges and Mean Sea-Level, Reprinted with Corrections*; John Wiley & Sons Ltd.: Chichester, UK, 1996; p. 486.

47. Cazenave, A.; Palanisamy, H.; Ablain, M. Contemporary sea level changes from satellite altimetry: What have we learned? What are the new challenges? *Adv. Space Res.* **2018**, *62*, 1639–1653. [CrossRef]

48. QUID forSea Level TAC DUACS Products. Quality Information Document, Copernicus Marine Environment Monitoring ServiceREF: CMEMS-SL-QUID-008-032-062, 10.09.2019, Issue: 2.1REF: CMEMS-SL-QUID-008-032-062, 10.09.2019, Issue: 2.1. 2019, p. 78. Available online: http://resources.marine.copernicus.eu/documents/QUID/CMEMS-SL-QUID-008-032-062.pdf (accessed on 29 February 2020).

49. AVISO+ Satellite Altimetry Data. Available online: https://www.aviso.altimetry.fr/en (accessed on 25 April 2019).

50. COPERNICUS—Marine Environment Monitoring Service. Available online: http://marine.copernicus.eu/ (accessed on 27 April 2019).

51. Woolf, D.; Tsimplis, M. The influence of the North Atlantic oscillation on Sea level in the Mediterranean and the Black Sea derived from satellite altimetry. In Proceedings of the Second International Conference on Oceanography of the Eastern Mediterranean and Black Sea: Similarities and Differences of Two Interconnected Basins, Ankara, Turkey, 14–18 October 2002; pp. 145–150. Available online: https://www.noc.ac.uk/publication/115146 (accessed on 27 February 2020).

52. Permanent Service for Mean Sea Level. Available online: http://www.psmsl.org/ (accessed on 14 April 2019).

53. Global Sea Level Observing System. Available online: https://www.gloss-sealevel.org/ (accessed on 14 April 2019).

54. General Directorate of Mapping. Turkish National Sea Level Monitoring System. Available online: https://tudes.harita.gov.tr/tudesportal/AnaEkranEN.aspx (accessed on 13 April 2019).

55. Nevada Geodetic Laboratory. Available online: http://geodesy.unr.edu/ (accessed on 19 June 2019).

56. General Directorate of Mapping. Turkish National Permanent RTK Network (CORS-TR/TUSAGA-Aktif). Available online: https://www.harita.gov.tr/english/u-13-turkish-national-permanent-rtk-network--tnpgn-active--and-determination-of-the-datum-transformations-parameters.html (accessed on 16 June 2019).

57. CATS. CATS Cycles Analysis and Timeseries Software, Version 1.0, Ray Tomes and Radiance Trust. 2010, p. 30. Available online: https://cyclesresearchinstitute.org/cats/ (accessed on 18 March 2020).

58. Avsar, N.B.; Kutoglu, S.H. Analysis of Seasonal Cycle of Sea Level Variations in the Black Sea. In Proceedings of the International Symposium on Applied Geoinformatics (ISAG-2019), Istanbul, Turkey, 7–9 November 2019; Bayram, B., Ed.; pp. 185–188. Available online: https://www.kongresistemi.com/panel/UserUploads/Files/c42d925e45daf45.pdf (accessed on 27 February 2020).

59. Korotaev, G.; Oguz, T.; Nikiforov, A.; Koblinsky, C. Seasonal, interannual, and mesoscale variability of the Black Sea upper layer circulation derived from altimeter data. *J. Geophys. Res. Ocean.* **2003**, *108*, 3122. [CrossRef]

60. Yildiz, H.; Andersen, O.B.; Kilicoglu, A.; Simav, M.; Lenk, O. Sea Level Variations in the Black Sea for 1993–2007 Period from GRACE, Altimetry and Tide gauge Data. In Proceedings of the EGU General Assembly 2008, Vienna, Austria, 13–18 April 2008. Available online: https://www.psmsl.org/about_us/news/2008/egu_2008/EGU2008-A-08684.pdf (accessed on 27 February 2020).

61. Legeais, J.-F.; Ablain, M.; Zawadzki, L.; Zuo, H.; Johannessen, J.A.; Scharffenberg, M.G.; Fenoglio-Marc, L.; Fernandes, M.J.; Andersen, O.B.; Rudenko, S. An improved and homogeneous altimeter sea level record from the ESA climate change initiative. *Earth Syst. Sci. Data* **2018**, *10*, 281–301. [CrossRef]

62. Avsar, N.B.; Jin, S.; Kutoglu, S.H. Interannual variations of sea surface temperature in the black sea. In Proceedings of the 2018 IEEE International Geoscience and Remote Sensing Symposium (IGARSS 2018), Valencia, Spain, 22–27 July 2018; pp. 5617–5620. [CrossRef]

63. Vigo, I.; Garcia, D.; Chao, B.F. Change of sea level trend in the Mediterranean and Black Seas. *J. Mar. Res.* **2005**, *63*, 1085–1100. [CrossRef]

64. Tsimplis, M.N.; Josey, S.A.; Rixen, M.; Stanev, E.V. On the forcing of sea level in the Black Sea. *J. Geophys. Res.* **2004**, *109*. [CrossRef]

65. Fenoglio-Marc, L.; Tel, E. Coastal and global sea level. *J. Geodyn.* **2010**, *49*, 151–160. [CrossRef]

 International Journal of
Geo-Information

Article

Cartographic Symbology for Crisis Mapping: A Comparative Study [†]

Ana Kuveždić Divjak *, Almin Đapo and Boško Pribičević

Faculty of Geodesy, University of Zagreb, 10000 Zagreb, Croatia; almin.dapo@geof.unizg.hr (A.Đ.); bosko.pribicevic@geof.unizg.hr (B.P.)
* Correspondence: ana.kuvezdic.divjak@geof.unizg.hr
† extended conference paper from GI4DM 2019.

Received: 23 January 2020; Accepted: 27 February 2020; Published: 28 February 2020

Abstract: Cartographic symbols on crisis maps serve as means of depicting information about the position, properties, and/or numerical values of objects, phenomena or actions specific to crisis mapping. Many crisis cartographic visualisations require simple, clear, categorised and visually organised symbols that can be easily read and understood by a wide range of crisis map users. Cartographic symbol sets for crisis mapping depend on effective graphic design, good availability (sharing and promotion, dissemination and promulgation) and standardisation (ensuring the general and repeatable use of map symbols). In this research, our aim was to examine the extent of these challenges in current cartographic symbology for crisis mapping. Through a comparative study of prominent symbol sets, we analysed efforts invested so far and proposed future directions. The results of this study may be of assistance in understanding less unified or coherent symbologies currently in use, or in revising or amplifying existing sets for future publication.

Keywords: cartographic symbols; map symbology; crisis map; comparative analysis; taxonomy; graphic design; availability; promulgation; sharing; standardisation

---

1. Introduction

During crisis response it is critical to share and understand complex spatial, thematic, and temporal information in a timely, visual and compelling way. Cartography plays an important role in delivering reliable, understandable, attractive, user-friendly, visual information through maps [1]. A crisis map is a thematic map on which objects, phenomena or actions specific to crisis management are represented according to their importance and highlighted using appropriate cartographic symbols [1,2]. Cartographic symbols on such maps serve as means of depicting information about the position, properties, and/or numerical values of objects, phenomena or actions specific to the crisis event. They are essential for communication to heterogeneous audiences in the unique environment of a crisis characterised by the immediate risk of considerable loss and stress. Consequently, many crisis cartographic visualisations rely on simple, clear, aesthetically pleasing symbols that can be read and understood easily by a wide range of crisis map users [2–4]. If they are incomprehensible, illegible, ambiguous, unclassified, random, or lack hierarchical organisation and other important design characteristics, they may fail to convey the intended message and complicate cooperative crisis management strategies at local, regional and international levels.

The problem of ineffective mapping that fails to communicate messages during a crisis was identified following Hurricane Andrew (in the Bahamas and south-eastern coast of the USA in 1992) and Hurricane Fran (in the USA in 1996) [3], when retrograde research was conducted on how the maps produced during or immediately after these events were used. The same problem was identified after major tragedies such as the 9/11 terrorist attack (in the USA in 2001), the Christmas tsunami (on the coasts of Indonesia, Thailand, Sri Lanka, and India in 2004) and Hurricane Katrina (in the USA, 2005).

Immediately after these events, issues were pinpointed such as the lack of cartographic symbols for communication in crisis situations, and visually overloaded maps which reduced legibility and made essential crisis information difficult to understand [2]. The need for cartographic symbols specifically adapted for use on crisis maps was highlighted. As a result, cartographic symbol sets were specially designed for communication and action in crises and were promoted within the crisis community.

Examples include (Figure 1):

- *Emergency Response Symbology* (*Homeland Security Working Group* (HSWG), *Federal Geographic Data Committee* (FGDC), USA, 2005) [5]
- *Canadian All-Hazards Symbology For Emergency Management* [6] (*Government Operations Centre Geomatics* (GOC), Canada, 2015) and its predecessors: *Canadian Disaster Database Symbology* (2007) and *Emergency Mapping Symbology* (2010)
- *Australian All Hazards Symbology* (*Emergency Management Spatial Information Australia* (EMSINA), Australia, 2007) and a revised edition issued in 2018 [7]
- *OCHA's Humanitarian Icons* (*United Nations Office for the Coordination of Humanitarian Affairs* (OCHA), International, 2012) and a completely revamped set of symbols released in 2018 [8]
- *MIL-STD-2525C Common Warfighting Symbology, Appendix G* (*Department of Defense* (DOD), USA, 2008) [9]
- *Humanitarian Demining Symbols* (*Geneva International Center for Humanitarian Demining* (GICHG), International, 2005) [10]
- *Symbol System for Disaster Management* (*Laboratory on Cartography, University of Architecture, Civil Engineering and Geodesy*, 2017, Bulgaria) [11]
- *European Emergency Symbology reference for 2D/3D maps* (*INDIGO project*, Europe, 2012) [12]
- *Civil Protection Common Map Symbology* (Ordnance Survey, UK, 2012) [13].

Despite the fact they were designed for the same purpose, these symbol sets differ in various aspects. At first glance, the most noticeable difference is the graphic appearance of the symbols—from economically stylized specific pictorial symbols to extremely simple geometric abstract forms (Figure 1). They also differ in how they are ordered, structured and visually organised within the set, how they are disseminated, promulgated and shared between organisations, the extent to which they are commonly recognised and accepted, etc.

Although the latest research in crisis management has mostly covered the technological aspects of improving efficiency and strengthening crisis response capacity [4], there is still a need for research examining characteristics important for determining how easy symbols are to use. Four general challenges related to the development of symbology for crisis mapping have been identified in a recent study by Kostelnick and Hoeniges [14] through a review of the cartographic literature and results of a survey conducted among different humanitarian organisations. These are: (1) symbol taxonomies, (2) symbol design issues, (3) symbol availability, sharing and promulgation, (4) standardisation process in the wider community.

In this research, our aim was to examine the extent of these challenges within current cartographic symbology for crisis mapping. Through a comparative study of prominent examples of existing symbol sets, our objectives were to analyse the efforts undertaken so far and propose future directions. We paid particular attention to sets that have undergone new, revised or amplified editions and examined whether the latest changes implemented succeed in meeting the challenges of crisis mapping symbology.

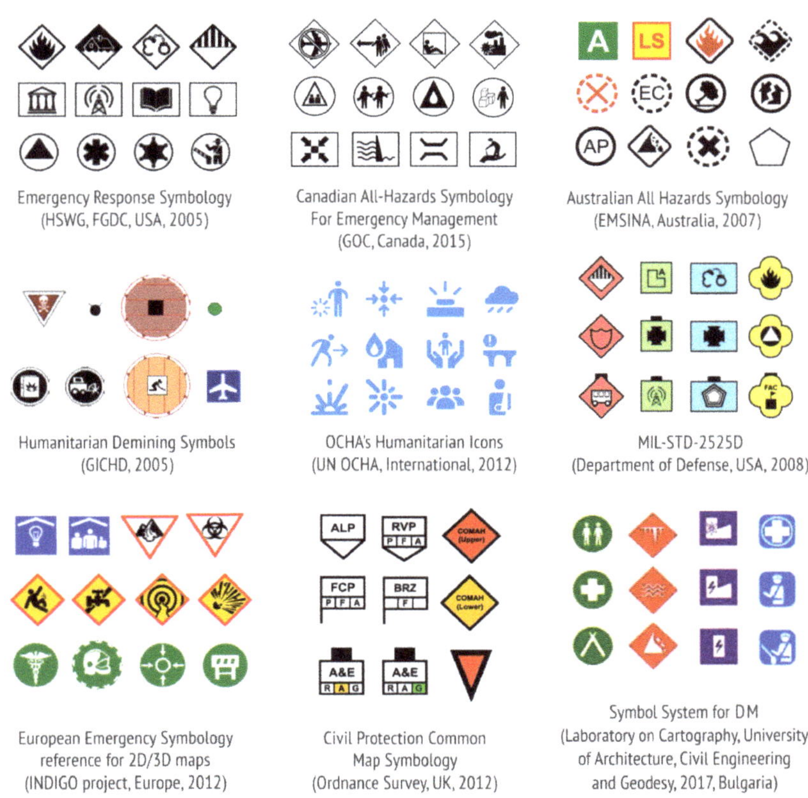

Figure 1. Examples of cartographic symbology for crisis mapping.

We were guided by the following research questions. *(1) Symbol taxonomies*: What does the taxonomy of cartographic symbols in sets and their internal breakdown look like? What graphic variables are used to support the visual and cognitive organisation of the symbols within the set? *(2) Symbol design*: Are all the basic geometric–graphical elements (point, line, and area) for depicting the position and quality of objects included in the set? Are the graphic and semantic qualities (such as simplicity, clarity, visibility, aesthetic appeal, traditionalism, familiarity, general acceptability, hierarchy, concreteness, semantic closeness) respected in their design? *(3) Availability*: Where can the symbol set be found and how easily? How and in which format have the symbols been shared? How are they promoted? Are there any study and training resources available (user manuals, best practice guidelines, etc.)? *(4) Standardisation*: Can the set be expanded with additional symbols? Are guidelines for the graphic design of new symbols provided? Has any assessment of the design, efficiency, and recognisability of cartographic symbols on crisis maps been carried out?

2. Materials and Methods

We examined six cartographic symbol sets published in different countries. Three were designed exclusively for crisis management (*American Emergency Response Symbology*, *Canadian All-Hazard Symbology*, *Australian All Hazard Symbology*), while two were intended for humanitarian activities (*OCHA's Humanitarian Icons* and *Humanitarian Demining Map Symbols*), and one for military operations (*MIL-STD-2525 Common Warfighting Symbology*). The main selection criterion was that all the cartographic symbol sets should contain symbols representing objects, phenomena and actions specific to crisis management, regardless of their primary purpose. Other criteria were their availability in the public domain and their recognition by the cartographic scientific and crisis mapping community [11,15–17].

To begin with, we systematised general information about each cartographic symbol set: its official name and country of origin, the responsible institution, the year it became publicly available, the chronology of addenda and/or new editions, the internet source, the format in which the symbols are available for download, and the terms of use.

We then conducted a comparative analysis to examine, compare and contrast specific aspects of symbol taxonomies, their design, standardisation and availability within the six selected cartographic symbol sets.

Regarding *(1) symbol taxonomies*, if the symbols in the set were classified in groups, we analysed comparatively their division into categories, pattern of arrangement, and order of connection. We analysed how thematic organisation in categories was transferred to the graphic appearance of the symbols. Transcription in a cartographic symbol set must be selective to clearly distinguish affiliation to a particular type, but also, within each type, it must be associative to clearly show its affiliation [18]. The available graphical variables (also known as visual variables outlined by Bertin [19]) are size, shape, colour hue, colour value, texture, and orientation. This initial set was later extended to include two variables used in cartographic design (colour saturation and arrangement) and three which are easier to manipulate through digital production methods (crispness, resolution, and transparency) [18]. Each visual variable can be used when designing appropriate cartographic symbols to show the position and quality of a discrete object, or to present information on its properties. However, each visual variable has specific properties and is more suitable for transcribing one aspect of information than others [19,20]. The criteria for the respective perceptual characteristics of visual variables and their semiotic association were based on findings documented in cartographic textbooks [18–21].

Regarding *(2) symbol design* we made a quantitative analysis of the total number of cartographic symbols in each set and their representation according to dimensions. Considering *dimensions*, the cartographic symbols were divided into point, line, area, volume, and space-time (four-dimensional) (according to [18]). We then analysed comparatively the implementation of four semantic qualities (concreteness, semantic closeness, familiarity, acceptability) and four graphic qualities (simplicity, visibility, consistency and aesthetic appeal) in the design of pictorial symbols. The choice of symbol qualities for analysis and the criteria for rating them were based on the general qualities of symbols documented in cartographic and behavioural literature through research into the characteristics important in determining how easy pictorial symbols are to use [22,23].

In terms of *(3) availability*, we analysed comparatively methods of sharing and promoting, disseminating and promulgating the cartographic symbol sets. We identified methods for disseminating symbols from existing sets, such as advertising, publications, presentations, workshops, brochures, flyers, posters, websites, exhibitions, conferences, training activities, innovation networks, and so on. We examined the technical aspects of how the cartographic symbols were shared, such as the format available for download, and whether they were embedded in existing GIS software (ArcGIS and QGIS) or symbol sharing platforms. We also researched and listed any available accompanying resources such as study and training materials, demonstrations of the use of symbols on maps, user manuals, best practice guidelines, etc.

In terms of *(4) standardisation*, we analysed comparatively measures taken regarding the general and repeated use of cartographic symbols from the set. We asked whether it was possible to extend the set with additional symbols, and whether there were any guidelines, requirements, rules for graphic design, or rules for implementing the symbols on crisis maps. Had an assessment of the design, efficiency, and recognition of cartographic symbols on crisis maps been carried out? Were there any recorded uses of the symbols on maps in real-case scenarios? We analysed whether the symbols were intended for use on a certain type of map at a certain scale.

3. Results

The elements of the cartographic symbols sets covered by the analysis are summarized and presented in Table S1 and given in the Supplementary Materials. The results of the comparative analysis, with specific examples, are reported in the subsections which follow.

3.1. Taxonomy, Visual and Hierarchical Organisation of Cartographic Symbols

A comparative analysis of the sets revealed different approaches to the hierarchical, thematic and visual organisation of cartographic symbols. For example, the symbols in *Canadian All-Hazards Symbology* and *Humanitarian Demining Map Symbols* were organised in three categories, while those in *American Emergency Response Symbology*, *Australian All Hazard Symbology* and *MIL-STD-2525 Common Warfighting Symbology* were organised in four categories. In the *OCHA's Humanitarian Icons*, many themes of interest to the humanitarian community, from natural disasters (such as tsunamis and earthquakes) to relief supplies (such as water containers and shelter kits) were covered. Complex humanitarian topics such as *access to people in need* and *protection of civilians* were also covered. However, no clear thematic division into categories was stated.

Although the total number of categories and their names in the sets differed, general similarities could be found. *Incidents*, *operations* and *infrastructure* are three commonly used categories for the thematic organisation of cartographic symbols for communication and action in a crisis. In *American Emergency Response Symbology*, *Canadian All-Hazards Symbology* and *Australian All Hazard Symbology*, visual organisation is achieved by connecting particular geometric shapes to particular categories of symbols (Figure 2a). In a new version of the *Australian All Hazard Symbology* (2018), a new category of observations has been added for features which are affected or impacted by the incident (Figure 2a).

In *American Emergency Response Symbology* a visual hierarchical status on the damage caused, marked by the particular geometric shape and/or colour hue of the symbol frame, can be additionally assigned to symbols in the operations and infrastructure categories (Figure 2b).

In the new edition of *Australian All Hazard Symbology*, graphic variables have been introduced for expressing ordered (hierarchical) properties (Figure 2c). A visual hierarchical status can also be assigned for features which are affected or impacted by the incident (Figure 2d).

Following the example of *American Emergency Mapping Symbology*, a new version of *Canadian All-Hazards Symbology* incorporates the use of different frames—a diamond for an incident, a rectangle for infrastructure, and a circle for operations. A frame with dashes represents a disruption to an incident or infrastructure. When the symbology set is distributed, these frames will be provided so users can combine them with any symbol [6].

In *MIL-STD-2525 Common Warfighting Symbology*, affiliation is shown by framing the symbols using different shapes and colour fills (for example, the relationship between an operator and an operative object). The basic categories of affiliation are *unknown*, *friendly*, *neutral* and *hostile* (Figure 2e).

Although the symbols in *OCHA's Humanitarian Icons* and *Humanitarian Demining Map Symbols* are thematically organised in categories, this has not been transferred to the graphic appearance of the symbols (Figure 2f). Since all the symbols in *OCHA's Humanitarian Icons* set are the same colour hue, associative and selective properties are not achieved. Although the pictograms in *Humanitarian Demining Map Symbols* use frames of different geometric shapes and colour fills, these variables are not applied to achieve visual organisation of the symbols, but arbitrarily.

3.2. Design of Cartographic Symbols

3.2.1. Representation of Cartographic Symbols according to Their Dimensions

A quantitative analysis of the cartographic symbols according to their dimensions showed that in the cartographic symbol sets, there were generally no line and area symbols envisaged for representing objects, phenomena, and actions specific to crisis management (see Table S1). The exceptions were the

Australian All Hazard Symbology (with 13 line and 10 area symbols), and the *Humanitarian Demining Map Symbols* (with 31 area symbols) (Figure 3).

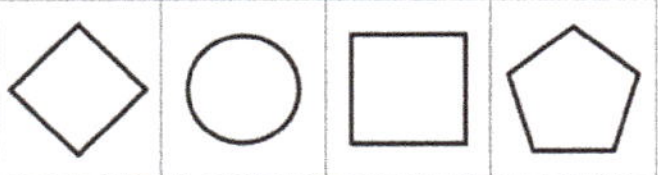

a) Incidents (diamond), Operations (circle) and Infrastructure (rectangle) are used for thematic and visual organisation of cartographic symbols in American Emergency Response Symbology, Canadian All-Hazards Symbology and Australian All Hazard Symbology. Category Observations (pentagon) is added in the new version of the Australian All Hazard Symbology.

b) In American Emergency Response Symbology a visual hierarchical status on the damage and operational level from Fully operational (left) to Totally incapacitated (right) can be additionally assigned to symbols in the categories Operations and Infrastructure.

c) Ordered property for representing the status of the asset Potentially Defendable (circle), Defendable (checkmark) and Not Defendable (cross) within the symbol frame is available in the new version of the Australian All Hazard Symbology.

d) In Australian All Hazard Symbology a visual hierarchical status on the damage for features which are affected or impacted by the incident from Total damage (left) to No damage (right) can be additionally assigned to symbols in the category Observations.

e) In MIL-STD-2525D Common Warfighting Symbology a yellow quatrefoil frame is used to denote unknown affiliation, a blue rectangle frame to denote friendly affiliation, a green square frame to denote neutral affiliation, and a red diamond frame to denote hostile affiliation.

f) Although OCHA's Humanitarian Icons cover many themes of interest to the humanitarian community, this thematic organisation has not been transferred to the graphic appearance of the symbols. The associative and selective properties of cartographic symbols are not achieved.

Figure 2. Taxonomy, visual and hierarchical organisation of cartographic symbols in the analysed symbol sets.

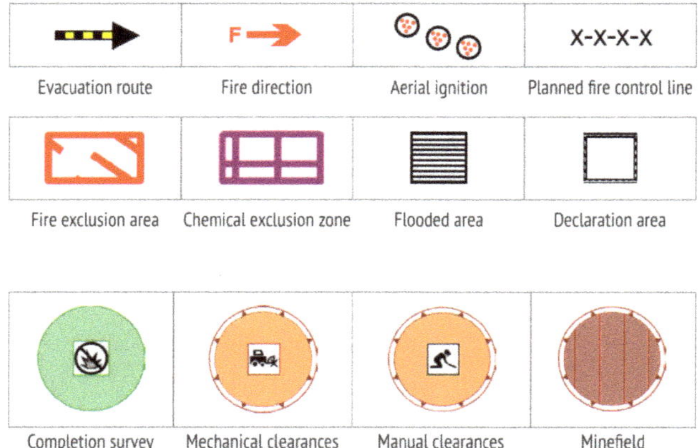

Figure 3. Examples of line (top) and area (middle) symbols from *Australian All Hazard Symbology,* and area symbols from *Humanitarian Demining Map Symbols* (bottom).

3.2.2. Concreteness of Cartographic Symbols

Concreteness refers to the extent to which the symbol depicts real objects, materials, or people [23]. Pictorial symbols tend to be more concrete and visually obvious in comparison to associative, geometric or alphanumeric versions [22].

For the comparative analysis of symbol concreteness in the sets, a quantitative analysis of representation of cartographic symbols according to appearance was made. Point cartographic symbols were divided into pictorial, associative, geometric and alphanumeric [18]. The symbols were considered *pictorial* if they were simplified drawings of important external (formal) or symbolic features of objects. The symbols were considered *associative* if a combination of geometric and pictorial characteristics was used to form a shape associated with the intended theme. If the symbols were derived from basic geometric shapes or from regular linear and area alignments of lines or points, they were regarded as *geometric*. *Alphanumeric symbols* were those formed using letters or numbers.

The quantitative analysis showed that associative, geometric and alphanumerical symbols were also represented in the sets in addition to pictorial ones. The exception was *OCHA's Humanitarian Icons* set where no geometric symbols were found. Examples of associative, geometric and alphanumeric symbols selected from each set are shown in Figure 4.

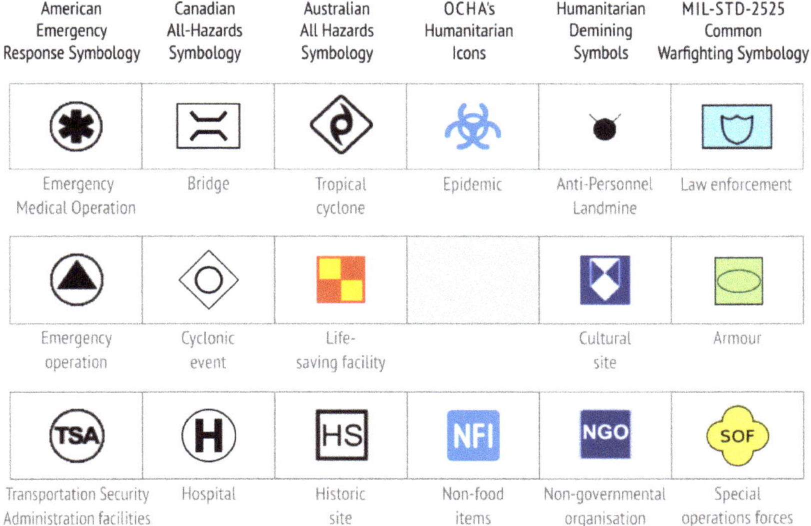

Figure 4. Selected examples of associative (top), geometric (middle) and alphanumeric (bottom) symbols in the analysed symbol sets.

3.2.3. Semantic Closeness of Cartographic Symbols

Pictorial symbols were further analysed on the basis of how the symbol (i.e. the visual representation) was linked to its meaning (i.e. the referent). Semantic closeness is the measure of the distance between the map symbol and what it is intended to represent [22,23].

At the general level, representation strategies in terms of visual similarity, semantic association and arbitrary convention were observed within the sets. Through visual similarity, the referent was represented by depicting its visual characteristics, as shown in selected pictograms from *OCHA's Humanitarian Icons* (Figure 5, top row). This kind of representation was regarded as the closest semantically.

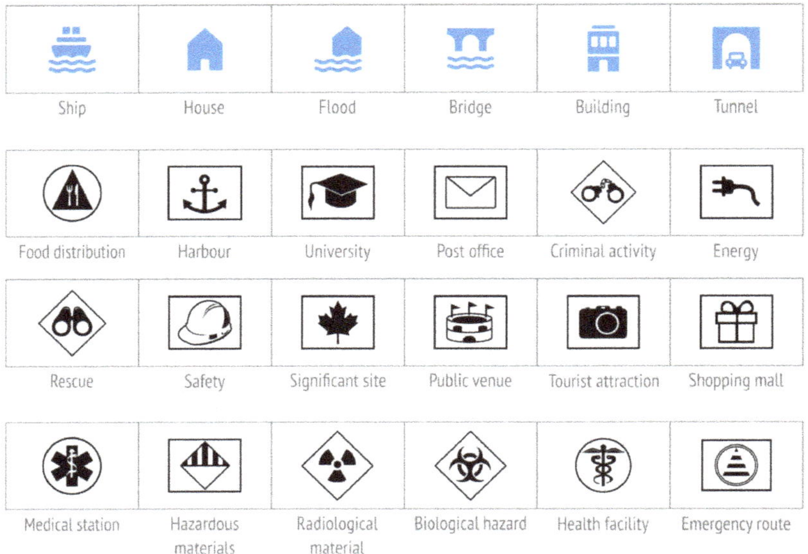

Figure 5. Examples of different representation strategies in the analysed symbol sets: visual similarity (top), semantic association (middle) and arbitrary convention (bottom). The representation is semantically closest on the top symbols, and semantically farthest on the bottom symbols.

Through semantic association, the connection between the referent and the symbol was mediated by depicting concepts that were semantically close to the referent. Semantic association in the selected pictograms from *Canadian All-Hazards Symbology* was achieved by depicting the visual characteristics of typical representatives (e.g. a knife and fork for a food distribution centre, a mortarboard for a university) (Figure 5, second row), or by depicting the visual characteristics of a pictogram with a higher level of meaning abstraction (e.g. binoculars for a rescue action, a helmet for safety, a maple leaf as the national symbol for a Canadian significant site) (Figure 5, third row).

The representation was semantically farthest when the referent was represented through an arbitrary convention that had to be learned in order to interpret it correctly (e.g. a rod of Asclepius for a health facility, or the international trefoil for radiological material) (Figure 5, bottom row).

3.2.4. Familiarity and Acceptability of Cartographic Symbols

Familiarity of symbols is the frequency with which they are encountered [22,23]. For example, most people find the symbols used to indicate public toilets very familiar, despite variations.

Cartographic symbols that have been used in almost unchanged or very similar forms over a long period of time were present in all the analysed sets (e.g. a knife and fork are usually used for a place where food is served, an envelope for a post office, an anchor for a ship port, along with commonly used cartographic symbols for a school, public recreation area, transport and bridge) (Figure 6).

Additionally, in *OCHA's Humanitarian Icon*, pictograms were identified that are regularly used in traffic communication, such as violent wind, snowfall, landslide and gas station, and to communicate public information, such as toilets and disabled persons. Generally accepted hazard pictograms were also present, e.g. for poison (which indicates danger in general, not just danger of poisoning), hazardous material, radioactive material and biohazardous infectious materials.

Figure 6. Selected examples of cartographic symbols that have been used in almost unchanged or very similar forms over a long period of time.

3.2.5. Visual Complexity (Simplicity) of Cartographic Symbols

The visual complexity of a symbol refers to the intricacy of its edges, the number of its elements, and level of detail [24]. A comparative analysis of cartographic symbols in the sets showed that the level of detail used to depict objects, phenomena and actions specific to crisis mapping differed, from economically stylized concrete pictorial symbols (e.g. *American Emergency Response Symbology* and *OCHA's Humanitarian Icons*) to extremely simple geometric abstract forms (e.g. *MIL-STD-2525 Common Warfighting Symbology*).

Cartographic symbols to depict civilians were selected and arranged from the simplest to the most detailed (Figure 7). In the first symbol, the person is represented by two lines and a circle. Not everyone will automatically associate this extreme simplification with a body and head. Some additional details such as lines indicating arms and legs may be necessary. On the other hand, too much detail can have the opposite effect. The examples in Figure 7 show that although an individual civilian symbol (from the old version of *Canadian All-Hazards Symbology*) works well on its own, a simpler pictogram like one from *American Emergency Response Symbology* works better for concepts like civil demonstrations, civilian displaced population, civil rioting and civilian evacuation.

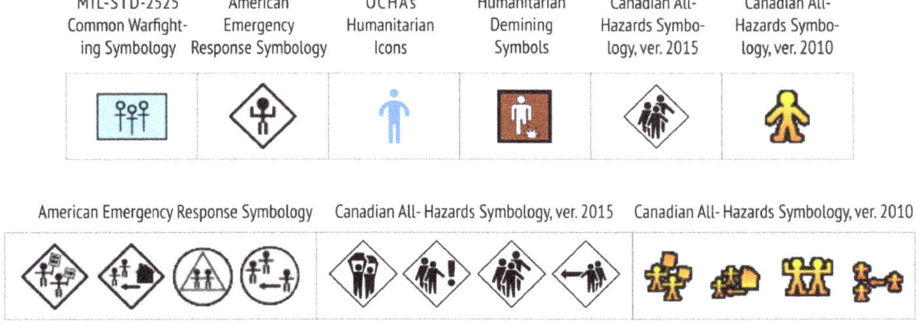

Figure 7. Different levels of visual complexity of cartographic symbols within the analysed sets.

3.2.6. Visibility of Cartographic Symbols

The visibility of a symbol refers to the ease with which it can be seen against the background [22]. In this research, we only analysed comparatively the *contrast* and *prominence* of pictograms against the colour inside the symbol (fill). Contrast of cartographic symbols against background maps should also be examined in the future research, to indicate how much the readability of each symbol locally decreases when it is presented on a specific colour background. A quantitative measure for colour contrast to be calculated between any map object and its background can be found in Reference [25].

Instructions for contrast ratings were obtained from previously calculated contrast differences between two colours, based on the light reflectancy readings in percentages for each of the two colours involved [26]. Good contrast can be achieved by using black and white [26]. Contrast in the form of a black pictogram on a white background was used for all symbols in *American Emergency Response Symbology*, *Canadian All-Hazards Symbology*, and some symbols in *Australian All-Hazard Symbology* and *Humanitarian Demining Map Symbols* (Figure 8).

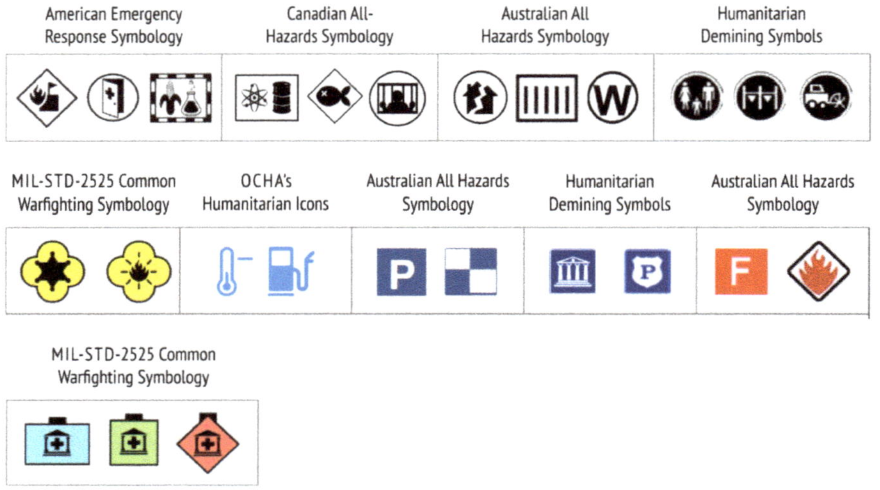

Figure 8. Examples of the contrast and prominence of pictograms set against the colour hue inside the symbol (fill) in the analysed sets.

In addition to light-dark contrast combinations [26], the following combinations of pictograms and backgrounds were deemed successful:

- black on yellow (symbols in *MIL-STD-2525 Common Warfighting Symbology* denoting the *unknown* affiliation category),
- red on white (certain symbols in *Australian All-Hazard Symbology*).
- blue on white and white on blue (all symbols in *OCHA's Humanitarian Icons*, and certain symbols in *Australian All-Hazard Symbology* and *Humanitarian Demining Map Symbols*).

Slightly lower contrast and prominence was noted for cartographic symbols in *MIL-STD-2525 Common Warfighting Symbology* denoting *friend*, *neutral*, and *hostile* affiliation categories. In these cases, black pictograms were used on blue, green and red backgrounds (Figure 8).

3.2.7. Consistency of Cartographic Symbols

Consistency means the extent to which symbols form a visually uniform set [22]. Consistency was analysed as a quality of the symbol set rather than as a quality of the individual symbols.

In all the analysed sets, pictograms were framed by particular geometric shapes and filled with predefined colour hues. Thus, a certain degree of consistency in the depiction of point symbols was achieved. The exception was the set of point symbols in *Humanitarian Demining Map Symbols*, where frames and colour fills were arbitrarily applied.

Consistency in a cartographic symbol set can also be achieved by using similar stroke weights, arcs, circle sizes, and perspectives [22]. The highest consistency was achieved in the new version of *OCHA's Humanitarian Icons*, where the new pictograms were drawn from scratch following predetermined standardized design rules. Thus, all the new icons looked similar in terms of visual complexity and appeared to belong to the same family.

More about the rules and guidelines for consistent and unified style and the visual appearance of cartographic symbols in the sets is discussed in Sections 3.4.2 and 3.4.3.

3.3. Availability (Sharing, Dissemination, and Promulgation) of the Cartographic Symbols

All the cartographic symbol sets examined are publically available online on the websites of the institutions responsible for their development. *American Emergency Response Symbology* has been publicly available on the web pages of the *Homeland Security Working Group of the Federal Geographic Data Committee* [27] since 2004. The current (third) version of *Canadian All-Hazards Symbology* was publicly released by *Government Operations Centre Geomatics* (GOC) in the document [6] in 2015. It is currently available at [28], but a permanent host is required. In 2018, the second edition of *Australian All Hazards Symbology* was released. The symbol set, related documents and entire project history are available on the website of *Emergency Management Spatial Information Australia* [29]. The *OCHA's Humanitarian Icons* set has been publicly available on the *United Nations Office for the Coordination of Humanitarian Affairs* (OCHA) website since 2012 [8]. The second edition was released in 2018. There has been a publicly available report [10] with a corresponding set of *Humanitarian Demining Symbols* on the website of the *Geneva International Center* [30] since 2015.

Detailed information for each cartographic symbol set (the responsible institution, the year it became publicly available, the last update, a chronology of addenda and/or new editions, the internet source, formats in which the symbols are available for download, and terms of use) is systematized in Table S1.

A comparative analysis showed that the most common formats in which symbols in the sets are shared are the raster PNG, and vector SVG format. Technical resources also include predefined style files for ESRI's ArcGIS for all the analysed symbol sets and for QGIS (in the case of *OCHA's Humanitarian Icons* and *Australian All Hazards Symbology*) that can be loaded into standard mapping software to promote easy sharing within and between organisations.

The *OCHA's Humanitarian Icons* set is the only representative of cartographic symbols for crisis and humanitarian mapping within the *Noun Project* [31], a platform that offers a crowd-sourced collection of universally recognisable icons for visual communication.

Symbols from the *Emergency Response Symbology* set are built-in in *Symbol Store*, a visual-enabled, web-based interactive tool designed to help mapmakers share point symbols [32]. The initial idea behind *Symbol Store* was to allow users to browse symbols by keywords, category tags, and contributors and to facilitate the discovery, retrieval and sharing of map symbol sets between users. Symbol sets can be downloaded as ESRI Style Files so that they can easily be imported into new or current ArcGIS map projects.

Joint Military Symbology XML (JointMilSyML or JMSML) is an XML schema and associated instance data, designed to document the contents of MIL-STD 2525D and NATO STANAG APP-6(C). The Military Overlay is supplied as a project template for ArcGIS Pro and allows military standard symbols to be created quickly using and adapting existing feature templates, creating a military overlay with military standard symbols and sharing the overlay as either a static image or a web map [9]. It is hoped that future defence and intelligence systems will be engineered to take advantage of this technology, thus accelerating the delivery of new military symbology to war combatants, reflected in updates to these standards.

3.4. Standardisation (General and Repeated Use) of Cartographic Symbols

3.4.1. Standardisation of Map Symbology

At the moment, only one set of truly standardised emergency symbology exists. This is *Emergency Response Symbology* used in the United States and standardised by the *American National Standards Institute* [5].

After its public release, the cartographic symbols in *Emergency Response Symbology* triggered great interest among crisis management experts and emergency service workers. Various software producers wanted to include the cartographic symbols in their software, which would have increased their availability and consistent use on crisis management maps. However, since the symbols were adopted as a standard of the *American National Standards Institute*, their use involves the payment of a fee, which has put off many users who are still using the free version. The standardised set is used officially by emergency management and first responder communities at all levels of need (national, state, local and incident) in the United States [27].

MIL-STD-2525 Common Warfighting Symbology is the standard setting out rules and requirements for defining and displaying military operations, and all units of the US Department of Defense (DOD) have been obliged to implement it since 2008. It is also available to non-DOD entities (e.g. first responders, the United Nations, and multinational partners).

The equivalents of this standard are two NATO publications: Allied Procedural Publication APP-6A—Military Symbols for Land Based Systems and Allied Procedural Publication APP-6B—Joint Symbology (1998), in which graphic symbols for marking units, positions and control measures in tactical operations are defined. The content of NATO's publications and U.S. Department of Defense's MIL-STD-2525 standards are basically the same, but the latter has been developing faster, and therefore the analysis in this paper refers to that standard.

3.4.2. Standardisation of Usage

In *Emergency Response Symbology* it is stated that the symbols are intended for use on digital and paper maps at large and medium scales. It is not recommended to use the symbols on small-scale maps, but rather simplified versions or geometric shapes that indicate the symbol category [27]. In *Australian All Hazard Symbology* it is stated that the symbols are intended for use on paper and digital topographic maps and aerial images at small, medium and large scales. There was no more detailed standardisation of their use in the resources studied. However, a new element in the version released in 2018 is the inclusion of five scale-dependent symbols for facilities (*Fire-Fighting Facility*, *Ambulance Facility*, *State Emergency Service Facility*, *Life-Saving Facility* and *Police Facility*) for use at smaller scales. Also, for the new category *Observations* in which frame fills in different colour hues of the same intensity selectively outline information on damage caused, alternatives to the black and white variants of the symbols are also envisaged [29]. In *Canadian All-Hazards Symbology* it is stated that the symbols are primarily intended for desktop mapping, while still enabling effective web use [6]. The symbols in the *OCHA's Humanitarian Icons* set are intended for use in a wide range of OCHA humanitarian community information products, including maps, written reports, infographics and websites, while the symbols in *MIL-STD-2525 Common Warfighting Symbology* are intended for use in paper military topographic maps, digital military information systems, "graphics" and "working maps" [9]. The symbols in the *Humanitarian Demining Map Symbology* set are intended for use in topographic maps and aerial images in digital and paper form, at large, medium and small scales, and are specially adapted for use in the mine action information set (Information Management System for Mine Action IMSMA) distributed by the Geneva International Humanitarian Demining Centre [30].

3.4.3. Extending Sets with New Symbols

The *Homeland Security Working Group* responsible for the development of *American Emergency Response Symbology* points out that the set does not include all the symbols required to represent objects, phenomena, and crisis-specific actions. If a need for new symbols arises, they will try to incorporate them in an existing set, depending on the resources and capabilities available [27]. However, guidelines for extending an existing set with new symbols are not publicly available.

The current version of the *Australian All Hazard Symbology* set does not include all the cartographic symbols needed to manage different crises. The existing symbols are limited to actions in certain types of crisis and provide a base that will be extended in the future to meet the wider needs of national security and crisis management [29]. The 2010 version contained a total of 83 symbols. Between 2011 and 2017, EMSINA continued to collect information for a new and/or improved *Australian All Hazard Symbology*. A dedicated symbology officer was elected in 2015. This person, in collaboration with a small EMSINA working group, revised the method for collecting and approving symbols in 2016. The latest version of the 2018 set contains a total of 127 symbols. However, there are no guidelines for extending the existing set with new symbols, that is, none have been published in the available resources. However, a workflow for new symbology proposals [29] has been clearly stated.

The *OCHA's Humanitarian Icons* set is periodically extended with new symbols as necessary [8], and as other versions of the set have been released, major advances have been made to standardise guidelines for extension with new symbols. In 2018, OCHA released a completely revamped set of 295 symbols (and the number continues to increase) as the result of a long, meticulous redesign process. The first version in 2012 contained a total of 241 symbols. It grew organically as illustrations were developed to meet internal design needs, and the new series has been drawn from scratch following standardised design rules. The OCHA Graphics Stylebook [8] was also released, containing guidelines for establishing some rules for all designers, so that there is consistency across the icon family. Moreover, the original set has been extended to include new themes (for instance cash transfer) and individual icons have evolved to reflect changes that have occurred since 2012 (in technology, for example).

3.4.4. Assessment

Assessing the symbol design and recognisability in the *American Emergency Response Symbology* set was conducted in two ways, and the test methods and detailed results were published [2,27]. In the first case, the assessment of the appearance of each symbol was conducted by the *Homeland Security Working Group* during December 2003 and January 2004, in an on-line open-type survey in which various crisis management and emergency services volunteers participated. The survey results were published in a report on the website of the *Homeland Security Working Group* [27]. For each symbol in the set, the participants of the survey indicated whether they accepted or rejected its graphic design and short definition. Symbols that did not reach the 75% acceptance threshold were reviewed and redesigned (e.g. 11 symbols were not accepted in the category 'incidents', 7 in the category 'natural events', 4 in the category 'activity', and none in the category 'infrastructure'). The symbols that met the threshold were accepted as standards of the American National Standardization Institute ANSI INCITS 415-2006 Homeland Security Mapping Standard—Point Symbology for Emergency Management. In the same period a recognition test was conducted of 15 randomly selected symbols from the category Incidents and 13 from the category Operations [2]. Since there are no clear guidelines or norms to test the recognition of cartographic symbols for a crisis, the standard recommendations ANSI Z535.3 National Standard for Criteria for Safety Symbols which prescribes general criteria for the assessment and use of safety symbols indicating specific hazards were adopted [2]. Fifty Californian fire-fighters participated in the testing, and it was found that only 6 of the 28 symbols rated achieved the 85% recognition level prescribed by the standard.

An assessment of the symbol design of *Humanitarian Demining Map Symbols* was conducted on the initial version. Professional pyro-technicians participated in the testing, and their comments and feedback were taken into account when adjusting the symbols in newer versions of the system [17].

There is no evidence that the design, effectiveness, or recognisability of the proposed cartographic symbols was assessed for other symbol sets covered by the existing literature and other available resources.

4. Discussion

Several recommendations may be made following the comparative analysis of cartographic symbol sets for crisis mapping presented in this paper.

4.1. Recommendations and Best Practices for the Taxonomy, Visual and Hierarchical Organisation of Cartographic Symbols

For a thorough understanding of a cartographic symbol set, that is, to achieve the optimal map function for communicating information in a crisis, symbols should be formed following the appropriate organisational structure or taxonomy. Although the data to be displayed on a map is sometimes already provided for the cartographer in a proper organisational structure, no such structure exists in the case of data for communication and acting in a crisis. However, the analysis of the sets showed that some similarities were found in the organisation of cartographic symbols in groups in *Emergency Response Symbology, Canadian All-Hazards Symbology, Australian All Hazards Symbology* and *MIL-STD-2525 Common Warfighting Symbology*.

The visual organisation of the symbols in the set should be such that crisis management participants (both cartographers and map users) notice it immediately [15]. This can be achieved by using the appropriate colour hues and different shapes for framing cartographic symbols, as in the sets in *Emergency Response Symbology, Canadian All Hazard Symbology* and *MIL-STD-2525 Common Warfighting Symbology*.

In Figure 9 we describe a possible cognitive scheme for interpretation of cartographic symbols in *Emergency Response Symbology, Canadian All-Hazards Symbology, Australian All Hazards Symbology* and *MIL-STD-2525 Common Warfighting Symbology* according to pattern we observed in the comparative analysis of their visual and hierarchical organisation. It was based on the similar example schema for interpretation of symbols on a National Park Service map and schema for interpreting U.S. Interstate Highway numbers presented in Reference [18].

The same model could be applied to the sets in *OCHA's Humanitarian Icons* and *Humanitarian Demining Map Symbols* to customise them for crisis mapping. Following such a cognitive scheme, users apply logic in the interpretation of cartographic symbols on a crisis map, and this tells them that the graphic appearance of the symbol is divided into two parts: the frame, which is to a certain extent a constant part of the cognitive scheme (that is, it can take a finite number of geometric shapes and colour hues), and the pictogram (a variable part of the scheme that takes on a new form each time). Users visually and/or logically interpret various kinds of pictograms, where each shape is associated with a particular object, phenomenon or action. The frame around a pictogram may be a red square or a blue rectangle, and users subconsciously organise them into groups by applying the similarity principle—similar objects form a group. The frame location on the map indicates the position of the displayed object in relation to other objects on the map. Apart from quality, the objects identified can also be distinguished by their ordered properties. By analysing the existing cartographic symbol set, it was noted that ordered properties were not present in the first versions but were included in later editions of *Australian All Hazards Symbology* and *Canadian All Hazards Symbology*. As a result, for example, infrastructure objects were labelled as *destroyed* or *undamaged*, and roads as *passable* or *obstructed*.

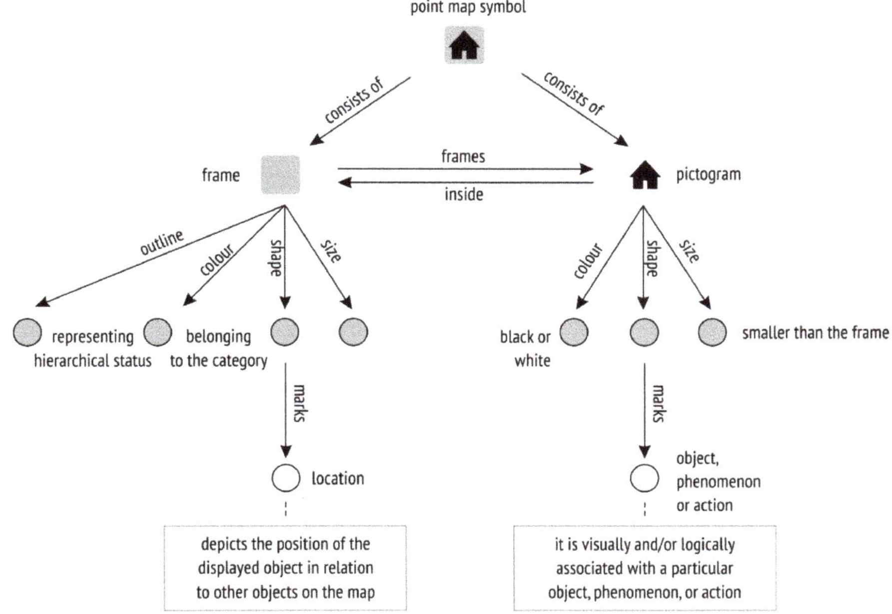

Figure 9. Cognitive scheme describing possible pattern of thinking and behaviour of users in the interpretation of cartographic symbols for crisis mapping.

4.2. Recommendations and Best Practices for the Design of Cartographic Symbols

Firstly, a quantitative assessment of existing symbols according to their dimensions indicated a lack of line and area symbols in *Emergency Response Symbology, Canadian All Hazard Symbology, OCHA's Humanitarian Icons, MIL-STD-2525 Common Warfighting Symbology* and lack of line symbols in *Humanitarian Demining Map Symbols*. It was apparent from the analysis of existing crisis maps [33] that line and area symbols were crucial in representing objects, phenomena, and actions specific to acting in a crisis, even though they appeared in much smaller numbers than point symbols. For example, line symbols were used to show fire front progression or oil spills at sea, and to mark evacuation routes or priority roads and routes used during a crisis. Area symbols usually indicated areas affected by crises (for example, flooded areas, wildfire-burned areas, or areas affected by hazardous gas leaks), and danger zones.

We are aware that line and area symbols in the cartographic symbol sets for crisis mapping are mapped directly on the base map, and that the length of a line symbol or shape and size of an area symbol cannot be defined in advance, as with point symbols. However, we recommend that these symbols are included in existing sets, with graphic variables applied instead of dimensions. These variables are size, direction, and colour hue (Figure 10). Size variation of the line (thicknesses or width) should be applied when designing the (hierarchical) and quantitative properties of objects. Direction variation (changing a line into an arrow) should be used to represent motion, and variation of colour value to express the ordered properties of line objects. When designing area symbols, two graphic variables should be applied: colour hue and transparency of fill and/or texture (Figure 10). The corresponding point symbol is located within the polygon and indicates which phenomenon is represented by the area symbol.

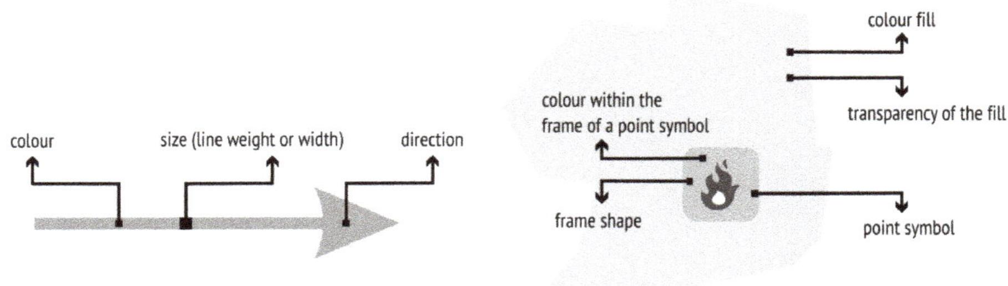

Figure 10. Integral geometric and graphic variables for depicting line and area symbols on crisis maps.

Line colour hues and shape fills should be applied consistently, according to the principle of similarity with the actual situation in nature, or by adopting symbolism. To achieve selective properties in the depiction of the line symbols, the thickness or width of lines must be at least 0.5 mm or 2 pixels apart [34]. The transparency of the area symbol fill can vary from 15 to 30 %, depending on the type of base map. The visual variable texture is well suited to associate map symbolisation with natural phenomena (e.g. snowflakes for snow avalanches). When used, it can denote categorical or numerical differences (the higher the value, the denser the texture). Sequential colour schemes are suited for ordered data that progress from low to high [19,21] and are should be used to depict classes of increasing values (e.g. blue value scale for intensities of the flooded areas). Diverging schemes put equal emphasis on mid-range critical values and extremes at both ends of the data range [19,21] and should be used for data whose values are above or below a critical value (e.g. a scale of red and green values for selective highlighting of danger zones).

Secondly, understanding the different semantic and graphic qualities of symbols is important when a new set of symbols is designed or an existing set extended with new symbols.

The quantitative assessment of existing symbols by frequency of appearance showed that although pictorial symbols were present to a greater extent, there were many examples of abstract and geometric symbols. In military systems such as *MIL-STD-2525 Common Warfighting Symbology* this is acceptable, since the users who need to interpret them have undergone special training. However, users who encounter these unfamiliar symbols for the first time may face difficulties. Thoughtfully designed pictorial symbols will not only be understood more easily and intuitively by a wide range of crisis participants such as crisis management experts and emergency services workers, but also by civilians or the general public. Therefore, appropriate pictorial symbols should be preferred over alphanumeric, geometric and abstract symbols. This requirement was considered in the new version of *Canadian All-Hazards Symbology*, where only one alphanumeric symbol was found (the letter H for *hospital*), whereas the previous version contained nine.

In terms of graphic qualities, when designing the visual appearance of the symbol, preference should be given to economically stylized, concrete pictorial symbols, such as those in *Emergency Response Symbology* and *OCHA's Humanitarian Icons*, compared to the more detailed ones in the old version of *Canadian All-Hazards Symbology*. However, extremely simple geometric abstract symbols, such as those in *MIL-STD-2525 Common Warfighting Symbology*, should be avoided.

The new version of *OCHA's Humanitarian Icons* is a successful example of the application of graphic and semantic qualities. In this symbol set, the pictogram is not treated as a single solution, but as a series of pictograms, taking into account concreteness, semantic closeness, familiarity, acceptability, simplicity, visibility, consistency and homogeneity.

4.3. Recommendations and Best Practices for the Availability of Cartographic Symbols

It is obvious that tradition, homogeneity, uniformity, and standardisation, both in the graphic design of symbols and their application on crisis maps, are crucial when creating map symbols for crisis mapping. The standardisation of cartographic symbols for crisis mapping (in the sense of ensuring unambiguous, consistent application) would allow users to become familiar with their meanings and increase their usefulness on crisis maps.

Emergency Response Symbology is arguably the most globally recognised standardised approach to emergency management mapping symbology and is also formally recognised as an American National Standards Institute (ANSI) standard. *Emergency Response Symbology* was the pioneer symbology standard for emergency management, and later attempts, including the *Canadian and Australian All-Hazards Symbology* sets, were inspired by and build as much as possible on it.

In addition to the graphic design of cartographic symbols, crisis management participants (cartographers and users) must be provided with rules and guidelines for use on the map. In order to expand the system with new symbols, guidelines for the graphic design of cartographic symbols must be standardised. Such guidelines must specify the minimum size for legibility and predict the use of symbols on maps at different scales, since scale dictates the size of a cartographic symbol and the amount of detail that can be represented by a pictogram in a particular symbol. We are aware that guidelines for determining the visual appearance of a particular symbol can only be general in nature, like guidelines for obtaining good legibility. So, those who design new symbols should have some (basic) knowledge of how to apply the guidelines.

Apart from the ease of understanding and memorising them, the success of cartographic symbols lies in their availability and maximum ease of use on future crisis maps. Incorporating symbols in software (e.g. the symbols in *American Emergency Response Symbology* are available in ESRI's ArcGIS software), and uploading them on platforms (e.g. the symbols in *OCHA's Humanitarian Icons* are available on the *Noun Project* platform) may help the set to be recognised as the *de facto* standard in the crisis and humanitarian community.

4.4. Recommendations and Best Practices for the Standardisation (General ad Repeated Use) of Cartographic Symbols

Since the current methods for public online sharing mostly include sharing via the organisation's website, future research in the field of crisis mapping should seek to develop additional resources (such as crowdsourced, open-source web-based repositories and platforms for accepting, storing and disseminating symbols) that would further encourage the sharing of symbol sets among organisations and promote standardisation with regard to ensuring unambiguousness and the general and repeated use of these symbols on crisis maps.

It is necessary to invest efforts in different forms of promotion, such as publishing, presentations, workshops, brochures, flyers, posters, conferences, and training activities. Sharing, promotion, dissemination and promulgation of cartographic symbols undoubtedly imply the costs of training, raising awareness, and changing standard practices and procedures. The establishment of funding mechanisms and a clear structure for the management of implementation activities should help mitigate these costs.

5. Conclusions

In this paper, a comparative analysis was conducted of six publicly available cartographic symbol sets that have been promoted since 2005 in the scientific cartographic and crisis management community. While future research could be extended to identify other existing symbol sets currently in use in professional and civilian crisis management, we believe that the limited number of six comparatively analysed cartographic symbol sets is justified in terms of making representative conclusions, mostly because of their wholeness, prominence and continuous history. A comprehensive analysis and parallel comparison of taxonomy, graphic design, availability (sharing, dissemination and promulgation) and standardisation (general and repeated use) of cartographic symbols resulted in an assessment of the current state of affairs, unresolved problems, and possible avenues of improvement regarding cartographic symbology for crisis mapping.

We recommend the inclusion of line and area symbols in existing sets, since they are often needed to depict evacuation routes during a crisis, or areas affected by the crisis. We propose a cognitive scheme with a recognised pattern of user thinking and behaviour in the interpretation of cartographic symbols for crisis mapping and advise the application of such a scheme to *OCHA's Humanitarian Icons* and *Humanitarian Demining Map Symbols* in terms of customisation for crisis mapping. Since appropriate pictorial symbols should always be preferred over alphanumeric, geometric and abstract symbols, we recommend the replacement of symbols in existing cartographic symbol sets with pictograph alternatives whenever possible. *OCHA's Humanitarian Icons* is a successful example of the application of consistency and homogeneity in symbol design. Following that example, we recommend that graphic and semantic qualities are considered in the design of cartographic symbols for crisis mapping. Regarding standardisation and efforts made in respect of the general and repeated use of cartographic symbols, we draw particular attention to *American Emergency Response Symbology* and *OCHA's Humanitarian Icons*. The availability of the symbols in *OCHA's Humanitarian Icons* within the *Noun Project* platform has undoubtedly helped this set to be recognised as the *de facto* standard in the crisis and humanitarian community.

Additionally, the comparative analysis showed that certain changes have been implemented in new, reviewed or extended editions of existing sets. Better visual organisation has been achieved in *Canadian All Hazards Symbology*, special symbols for expressing associative and selective properties have been added to *Canadian* and *Australian All Hazards Symbology*. Study and training materials such as demonstrations of using symbols on maps have been provided with *Emergency Response Symbology* and *Canadian All-Hazards Symbology*, and graphic guidelines have been produced for extending the *OCHA's Humanitarian Icons* set.

We hope that the results of this comparative study of prominent cartographic symbols for crisis mapping will be of assistance in understanding less unified and coherent symbologies currently in use and in the production of future revisions or amplifications of existing systems. Good practices implemented in existing cartographic symbol sets for crisis mapping can also reduce duplicated efforts and encourage the adoption of existing symbol sets.

Although compliance with the guidelines, rules, and graphic requirements outlined in this comparative analysis can be used as criteria for assessing the appearance, effectiveness and visibility of the existing cartographic symbology for crisis mapping, such process will provide partly subjective and partly objective assessment of the symbol qualities. While some conditions, such as the assessment of the applied basic geometric and graphic variables, visibility, concreteness and symbol dimensions can be accurately measured, other conditions, such as the assessment of simplicity, familiarity and semantic closeness of cartographic symbols, will depend on the subjective impression, background, and abilities of the observer.

More research is needed with a focus on the empirical evaluation of the comprehension and usability of the existing symbology for crisis mapping. User-testing should include a heterogeneous audience of crisis map users. Perceptual aspects of map symbology on the various background maps (e.g. satellite and aerial imagery, topographic map, relief map, city map) should also be considered to better understand all user aspects of cartographic symbology for crisis mapping and the factors that influence them. Furthermore, eye-tracking methods can provide assistance for objective and quantitative evidence in cognitive research on cartography [35,36]. By recording real-time fixation, saccade, and duration data, and analyzing eye movement behavior they can provide more direct suggestions to users' visual cognition of crisis maps and help in developing improved methods for crisis mapping.

Supplementary Materials: Table S1: Results of a comparative study of six publicly available cartographic symbol sets for crisis mapping, Repository R1: Digital repository of cartographic symbols covered by the analysis.

Author Contributions: Conceptualization, Ana Kuveždić Divjak; Funding acquisition, Boško Pribičević; Methodology, Ana Kuveždić Divjak; Visualization, Ana Kuveždić Divjak; Writing—original draft, Ana Kuveždić Divjak; Writing—review & editing, Ana Kuveždić Divjak, Almin Đapo and Boško Pribičević All authors have read and agreed to the published version of the manuscript.

Acknowledgments: This research was fully supported by the Croatian Science Foundation, project number IP-01-2018-8944.

Conflicts of Interest: The authors declare no conflict of interest.

References

1. Bandrova, T.; Zlatanova, S.; Konecny, M. *Geoinformation for Disaster and Risk Management, Examples and Best Practices*; Joint Board of Geospatial Information Societies (JB GIS) and United Nations Office for Outer Space Affairs (UNOOSA): Copenhagen, Denmark, 2010; ISBN 978-87-90907-88-4.

2. Akella, M.T. First Responders and Crisis Map Symbols: Clarifying Communication. *Cartogr. Geogr. Inf. Sci.* **2009**, *36*, 19–28. [CrossRef]

3. Dymon, U.J. Mapping—The Missing Link in Reducing Risk under SARA III. *Risk* **1994**, *5*, 337.

4. Konecny, M.; Bandrova, T.; Kubicek, P.; Marinova, S.; Stampach, R.; Stachon, Z.; Reznik, T. Digital Earth for Disaster Mitigation. In *Manual of Digital Earth*; Guo, H., Goodchild, M., Annoni, A., Eds.; Springer: Singapore, 2020. [CrossRef]

5. ANSI INCITS-415 2006. *Homeland Security Mapping Standard—Point Symbology for Emergency Management*; American National Standard for Information Technology: Washington, DC, USA, 2006.

6. Government of Canada's Government Operations Centre: Canadian All-Hazards Symbology. Available online: https://www.publicsafety.gc.ca (accessed on 5 December 2019).

7. Emergency Management Spatial Information Australia: Australian All Hazards Symbology Version 2—2018. Available online: https://www.emsina.org/allhazardssymbology (accessed on 5 December 2019).

8. United Nations Office for the Coordination of Humanitarian Affairs. Humanitarian and Country Icons 2018. Available online: https://reliefweb.int/report/world/humanitarian-and-country-icons-2018 (accessed on 5 December 2019).

9. Department of Defense. *MIL-STD-2525D 10 June 2014, Interface Standard Joint Military Symbology*; Department of Defense: Washington, DC, USA, 2014.

10. Kostelnick, J. *Cartographic Recommendations for Humanitarian Demining Map Symbols in the Information Management System for Mine Action (IMSMA)*; Geneva International Centre for Humanitarian Demining: Geneva, Switzerland, 2005.

11. Marinova, S. New Map Symbol System for Disaster Management. *Proc. Int. Cartogr. Assoc.* **2018**, *1*, 74. [CrossRef]

12. INDIGO Project Crisis Management Solutions. European Emergency Symbology Reference for 2D/3D Maps Developed within the INDIGO Project. Available online: http://indigo.diginext.fr (accessed on 5 December 2019).

13. Cabinet Office with the Ministry of Defence and the Ordnance Survey. Civil Protection Common Map Symbology. Available online: https://www.ordnancesurvey.co.uk/support/symbols-for-emergencies.html (accessed on 5 December 2019).

14. Kostelnick, J.C.; Hoeniges, L.C. Map Symbols for Crisis Mapping: Challenges and Prospects. *Cartogr. J.* **2019**, *56*, 59–72. [CrossRef]

15. Bianchetti, R.A.; Wallgrün, J.O.; Yang, J.; Blanford, J.I.; Robinson, A.C.; Klippel, A. Free Classification of Canadian and American Emergency Management Map Symbol Standards. *Cartogr. J.* **2012**, *49*, 350–360. [CrossRef]

16. Robinson, A.C.; Roth, R.E.; MacEachern, A.M. Challenges for map symbol standardization in crisis management. In Proceedings of the 7th International ISCRAM Conference, Seattle, WA, USA, 2–5 May 2010.

17. Kostelnick, J.C.; Dobson, J.E.; Egbert, S.L.; Dunbar, M.D. Cartographic symbols for humanitarian demining. *Cartogr. J.* **2008**, *45*, 18–31. [CrossRef]

18. MacEachren, A.M. *How Maps Work: Representation, Visualization and Design*; Guilford: New York, NY, USA, 1995; ISBN 978-15-72300-40-8.

19. Bertin, J. *Semiology of Graphics: Diagrams, Networks, Maps*; University of Wisconsin Press: Madison, WI, USA, 2010; ISBN 978-15-89482-61-6.

20. Roth, R. Visual Variables. In *The International Encyclopedia of Geography*; Richardson, D., Castree, N., Goodchild, M.F., Kobayashki, A., Liu, W., Marston, R.A., Eds.; Wiley: Hoboken, NJ, USA, 2017; pp. 1–11. ISBN 978-0-470-65963-2.

21. Robinson, A.H.; Morrison, J.L.; Muehrcke, P.C.; Kimerling, A.J.; Guptill, S.C. *Elements of Cartography*, 6th ed.; John Wiley & Sons, Inc.: Hoboken, NJ, USA, 1995.

22. Korpi, J.; Ahonen-Rainio, P. Cultural Constraints in the Design of Pictographic Symbols. *Cartogr. J.* **2010**, *47*, 351–359. [CrossRef]

23. Mcdougall, S.J.P.; Curry, M.B.; de Bruijn, O. Measuring symbol and icon characteristics: Norms for concreteness, complexity, meaningfulness, familiarity, and semantic distance for 239 symbols. *Behav. Res. Methods Instrum. Comput.* **1999**, *31*, 487. [CrossRef] [PubMed]

24. Forsythe, A.; Sheehy, N.; Sawey, M. Measuring icon complexity: An automated analysis. *Behav. Res. Methods Instrum. Comput.* **2003**, *35*, 334. [CrossRef] [PubMed]

25. Lupa, M.; Szombara, S.; Chuchro, M.; Chrobak, T. Limits of Colour Perception in the Context of Minimum Dimensions in Digital Cartography. *ISPRS Int. J. Geo-Inf.* **2017**, *6*, 276. [CrossRef]

26. Arthur, P.; Passini, R. *Wayfinding-People, Signs, and Architecture*; McGraw-Hill: New York, NY, USA, 1992.

27. Homeland Security Working Group: Symbology Reference. Available online: https://www.fgdc.gov/HSWG/index.html (accessed on 5 December 2019).

28. Canadian All-Hazards Symbology. Available online: http://216.254.169.195/CAHS_SCTR (accessed on 5 December 2019).

29. Australasian All Hazards Symbology. Available online: https://www.emsina.org/allhazardssymbology (accessed on 5 December 2019).

30. Geneva International Centre for Humanitarian Demining. Available online: https://www.gichd.org (accessed on 5 December 2019).

31. The Noun Project. Available online: https://thenounproject.com/ (accessed on 5 December 2019).

32. Robinson, A.C.; Pezanowski, S.; Troedson, S.; Bianchetti, R.A.; Blanford, J.I.; Stevens, J.; Guidero, E.; Roth, R.E.; MacEachren, A.M. Symbolstore: A Web-Based Platform for Sharing Map Symbols. *Cartogr. Geogr. Inf. Sci.* **2013**, *40*, 415–426. [CrossRef]

33. Kuveždić Divjak, A.; Lapaine, M. Crisis Maps—Observed Shortcomings and Recommendations for Improvement. *ISPRS Int. J. Geo-Inf.* **2018**, *7*, 436. [CrossRef]

34. Spiess, A.; Baumgartner, U.; Arn, S.; Vez, C. *Topographic Maps—Map Graphic and Generalisation*; Cartographic Publication Series No. 17; Swiss Society of Cartography: Zurich, Switzerland, 2002.

35. Popelka, S.; Vondrakova, A.; Hujnakova, P. Eye-tracking Evaluation of Weather Web Maps. *ISPRS Int. J. Geo-Inf.* **2019**, *8*, 256. [CrossRef]

36. Brychtova, A.; Coltekin, A. An Empirical User Study for Measuring the Influence of Colour Distance and Font Size in Map Reading Using Eye Tracking. *Cartogr. J.* **2016**, *53*, 202–212. [CrossRef]

International Journal of *Geo-Information*

Article

Use of Mamdani Fuzzy Algorithm for Multi-Hazard Susceptibility Assessment in a Developing Urban Settlement (Mamak, Ankara, Turkey)

Tugce Yanar [1], Sultan Kocaman [1,*] and Candan Gokceoglu [2]

[1] Department of Geomatics Engineering, Hacettepe University, 06800 Beytepe Ankara, Turkey; tugceyanar@hacettepe.edu.tr

[2] Department of Geological Engineering, Hacettepe University, 06800 Beytepe Ankara, Turkey; cgokce@hacettepe.edu.tr

* Correspondence: sultankocaman@hacettepe.edu.tr

Received: 7 December 2019; Accepted: 17 February 2020; Published: 19 February 2020

Abstract: Urban areas may be affected by multiple hazards, and integrated hazard susceptibility maps are needed for suitable site selection and planning. Furthermore, geological–geotechnical parameters, construction costs, and the spatial distribution of existing infrastructure should be taken into account for this purpose. Up-to-date land-use and land-cover (LULC) maps, as well as natural hazard susceptibility maps, can be frequently obtained from high-resolution satellite sensors. In this study, an integrated hazard susceptibility assessment was performed for a developing urban settlement (Mamak District of Ankara City, Turkey) considering landslide and flood potential. The flood susceptibility map of Ankara City was produced in a previous study using modified analytical hierarchical process (M-AHP) approach. The landslide susceptibility map was produced using the logistic regression technique in this study. Sentinel-2 images were employed for generating LULC data with the random forest classification method. Topographical derivatives obtained from a high-resolution digital elevation model and lithological parameters were employed for the production of landslide susceptibility maps. For the integrated hazard susceptibility assessment, the Mamdani fuzzy algorithm was considered, and the results are discussed in the present study. The results demonstrate that multi-hazard susceptibility assessment maps for urban planning can be obtained by combining a set of expert-based and ensemble learning methods.

Keywords: multi-hazard; susceptibility mapping; developing urban settlements; landslide; flood; logistic regression; Mamdani fuzzy algorithm; M-AHP

---

1. Introduction

Improved disaster management is an important focus locally and globally to reduce the losses caused by natural disasters [1]. The harmful effects of natural hazards on human lives and economies increase with the inadequate land-use planning in developing countries. Actual infrastructure and land use should be taken into account in urban planning, which are often neglected [2]. Urban planning is a complex procedure that needs to consider existing infrastructure, human use, and natural hazards. The recent developments in geoinformation technologies for data collection and analysis, such as photogrammetry, remote sensing, three-dimensional (3D) geographical information systems (GIS), Web-GIS, volunteered geographical information (VGI), and advanced spatial analysis methods, can support this procedure and provide the essential tools to develop a combined approach.

Among the geology- and climate-related natural hazards (i.e., landslides, floods, earthquakes, droughts, wildfires, tornados, volcanic eruptions, and avalanches) [3], urban areas are mostly affected by landslides and floods. Landslide is one of the most common natural hazards with global spread, and it may damage buildings, infrastructure, and other facilities in urban areas [4]. Between 1950 and 2018, 23,041 landslides were observed in Turkey [5]. Although there is a vast amount of research on landslide susceptibility assessment in the literature (e.g., References [6–10]), most studies were on open lands and forests. Landslide susceptibility assessment in urban areas is difficult due to dense construction and buildings that modify and largely cover the topography. In addition, further research is needed due to the complexity of the problem, as well as incomplete and temporally inaccurate landslide inventories [11]. Production of regional landslide susceptibility maps can be difficult due to the requirement of actual data. Such maps are essential for urban planning and disaster mitigation efforts carried out by governments.

Although landslide conditioning parameters can be manifold, the main limitation for the number of factors employed to produce landslide susceptibility maps is the data availability. In general, geomorphological (e.g., topographical, hydrological, etc.), geological (e.g., lithology), and land-use and land-cover (LULC) parameters must be considered for this purpose [4]. A dense digital terrain model (DTM) can be used to characterize the geomorphology in detail in an urban area. In comparison to LULC, the geological and geomorphological characteristics change slowly. Since the LULC data are very important for landslide susceptibility assessment [12], actual data to demonstrate the LULC are required for obtaining high accuracy. In addition to the conditioning parameters, existing landslide inventories are employed in the susceptibility assessment models. On the other hand, landslide inventory extraction in settlement areas and heavy construction sites is also difficult due to covered or modified topography that considerably obstructs the visibility of landslides.

Floods also constitute one of the most commonly occurring and destructive natural hazards in the world, which also cause the highest number of fatalities [13,14]. According to the statistics [15], increasing numbers of flood events are being observed in Turkey. Depending on climate change and rapid urbanization, the number will continue to rise in the next decade in Turkey. Flood susceptibility maps also contain essential information for mitigation efforts, and they must be taken into account by decision-makers [16–18]. Flood susceptibility assessments for Ankara city were produced by Sozer et al. [19,20] using an expert-based decision support system called the modified analytical hierarchy process (M-AHP) [21]. A small part of the flood susceptibility map that covers the study area was employed here for multi-hazard susceptibility assessment. It should be noted that the flood susceptibility must be evaluated at a regional level, which includes the extent of the basin, since using the data of only a small area with limited altitude extent would lead to incomplete data and misinterpretation.

On the other hand, landslides and floods are often effective over the same regions since they occur in areas with similar geomorphological and climate conditions. Heavy precipitation triggers both floods and landslides, which occur one after the other. Therefore, areas that are prone to floods and landslides need to be assessed together to understand the combined effects of both.

The main aim of this study was to develop a spatial analysis methodology to predict the combined flood and landslide susceptibility level in a developing urban settlement, which can then be used as a basis for land-use planning. A part of Mamak District in Ankara, Turkey was selected as the study area for this purpose, because Mamak District is an unplanned settlement area of Ankara.

In the initial phase of the study, the landslide susceptibility map was produced by extracting and using the up-to-date LULC data from Sentinel-2 satellite images, high-resolution DTM, and lithology data obtained from existing geodatabases [4]. Since the flood susceptibility maps must be produced at regional level, i.e., basins, a part of the map produced by Sozer at al. [19,20] was reclassified and used for the purposes of this study. To assess the multi-hazard susceptibility level (MHSL), a Mamdani fuzzy algorithm was developed, and the results are presented and discussed in the later sections. This is the first application of the Mamdani fuzzy algorithm to combine two susceptibility maps in the international literature.

2. Background on Multi-Hazard Assessment

Many natural hazards types affect urban settlements. To mitigate the effects of future natural hazards, all possible risks in an area should be assessed. Multi-hazard assessment models can be generated by integrating multiple susceptibility assessment maps belonging to different types of natural hazards for a specific area. Moreover, using multi-layer information is a more reliable and effective approach to disaster prevention [22]. Different types of susceptibility maps created by using various parameters and factors can be combined with different methods, such as AHP, which is based on expert opinion. Using AHP, a suitability map can be determined by the weight coefficients and uncertainties for each hazard [23–25]. Furlan et al. [26] provided a gradual analysis of all components contributing to the risk at a particular site. This method involved the assessments of hazard, exposure, vulnerability, and risk. In this analysis, since the input dataset was large and heterogeneous, the multi-criteria decision analysis (MCDA) method was used to evaluate the parameters [27]. In all these studies, scores and weights provided by the experts had great importance, and it was stated by the researchers that they affected the accuracy of the results. Chen et al. [28] considered debris flows and river and flash flooding to be common in one area. These hazard types were examined in four scenarios (major, moderate, minor, and frequent events). In this approach, the losses caused by each hazard were firstly calculated individually. Afterward, the spatial probability of the element at risk, the physical vulnerability, and the quantification of the exposed elements at risk were multiplied. The effects of hazard types were compared based on the results.

It is also important to include social and economic dimensions in multi-hazard assessment methods [29]. China's disaster risk index was calculated for 31 provinces by using four types of factors: exposure (population exposed to earthquakes, floods, droughts, low temperatures/snow, and gale/hail), susceptibility (based on public infrastructure, income health, and economic status), coping capacity (based on governance, medical care, and material security), and adaptive capacity (related to future natural events) [30]. In addition to these factors, the exposure parameter was analyzed as hazard exposure and hazard loss. Hazard exposure referred to the presence of assets and values that may be adversely affected in hazardous areas. The hazard loss was defined by the extent of physical damage, monetary loss, human loss, and economic deterioration. At the same time, the scope of the other parameters was extended [31]. Using a multiple linear regression method, influencing factors of community resilience were calculated. It was also stated that this model can be developed for employing a different weighting scheme by using expert knowledge and the entropy. Moreover, for holistic multi-hazard assessment methodologies, it was proposed that anthropogenic processes should be used in relation to natural disasters, and it was mentioned that the environment in which natural hazards are experienced is shaped by human activities. The main idea was to include this relationship in the multi-hazard assessment process [32,33]. Additionally, Gallina et al. [33] and Basheer Ahammed and Pandey [22] pointed out that the climate change perspective is a forgotten piece that should be evaluated in a multi-risk assessment of natural and anthropogenic systems.

Barrantes [34] proposed a natural multi-hazard assessment model that can be used when working with limited data. This model proposed an algorithm that takes the spatial overlap of the values of each risk and the potential interaction between different natural risks and the temporal frequencies into account. In this algorithm, all risk combinations were evaluated and potential interactions between natural hazards matrix were formed. Potential multi-hazard risks arose with the set of intersections of all combinations. Liu et al. [35] used the MmhRisk-HI (Model for multi-hazard Risk assessment with a consideration of Hazard Interaction) method for multi-hazard risk assessment. The method had two main components. The first component analyzed the relationship between the hazardous environment and the hazards, showing the probability of multiple hazard occurrence. In this component, the probabilities were calculated with functions used according to the relationship levels (independent, mutex, parallel, series) between the natural hazards. The second component calculated the possible damages and loss rate by employing a Bayesian network. In Bernal et al. [36], a fully probabilistic multi-hazard risk model was assessed for hazard, exposure, vulnerability, and loss using specific stochastic event sets. The average and maximum yearly losses were calculated with this model. Quantitative approaches were also used in some regional risk assessments [37]. The study of quantitative multi-hazard risk evaluation using elements at risk, their exposure, and their vulnerability can be examples for regional multi-hazard assessment. In another quantification approach, the interrelation between hazards was examined in two ways: by cascading hazards and by compound hazards. These relationships were evaluated with three hazard interrelation modeling approaches (stochastic, empirical, mechanistic) [38]. When the results were evaluated using different model parameters in each approach, it was seen that the extreme copulas method in the stochastic approach, the linear regression method in the empirical approach, and hydrodynamic models in the mechanistic approach were most prevalent. As mentioned, the use of mechanistic and stochastic methods in multi-parameter (more than two) hazards imposes certain restrictions, such as uncertainties caused by statistical assumptions (e.g., distribution, dependence model selection) or the effect of the data quality used for validating the mechanistic models [38].

Assessing multiple hazards in urban areas and predicting future risks can help decision-makers to prioritize actions and manage the risks [32,39–41]. Although the effects of natural risks on the urban area were examined individually and the results were compared visually, a quantitative approach of the co-evaluation process of hazards was lacking in Chang et al. [39] and Jacobs et al. [42]. A combined and quantitative assessment of hazards provides more accurate results than individual assessments and visual comparison. However, the choice of parameters and their weights, as well as the quality of data, are also important. The choice of using a qualitative, semi-quantitative, or quantitative approach may vary depending on the target [43]. In Omidvar and Karimi [44], a method developed using the theory of probability and Boolean logic was used. This study was conducted for multi-hazard reliability measurements according to available urban data. Multi-hazard reliability emerged with operations on different combinations of hazard risks.

Machine learning methods are also available for multi-hazard risk assessment. In Reference [45], risk analysis for each natural hazard type was carried out by modifying the weights of each parameter according to the hazard type using the random forest (RF) method. Afterward, a multi-hazard map was produced by combining the results. The model accuracy of 96.70% supported the usefulness of machine learning in multi-hazard risk assessments. Mirzaei et al. [46] used Technique for Order of Preference by Similarity to Ideal Solution (TOPSIS) as a multi-criteria decision-making model. It worked with similarity indexes of hazard maps and showed multi-hazard assessment. In Sheikh et al. [47], TOPSIS–Mahalanobis distance, TOPSIS, and simple additive weight (SAW) methods were combined. Although the TOPSIS method was criticized for using only a geometric distance, it was mentioned that it gives more clear results for natural hazards.

There are also other studies in the literature in which different multi-criteria decision-making methods were used in risk assessment. In Pourghasemi et al. [48], risk maps were created by using the stepwise weight assessment ratio analysis (SWARA) method which is an expert-oriented method, as well as the adaptive neuro-fuzzy inference system (ANFIS) method which involved an FIS (fuzzy inference system) and gray wolf optimization (GWO) to find the optimal solution. In multi-hazard analysis, the occurrence rate of hazard risk combinations was determined using weighted overlay analysis of risk maps by Mukhopadhyay et al. [27]. Moreover, with the use of fuzzy modeling, direct standardization of multiple indicators, aggregation, and deriving the impact are possible. By using the gamma fuzzy overlay model, the relationship between multiple input criteria was explored [49]. Kappes et al. [50] emphasized the importance of multi-hazard assessment and compiled difficulties when analyzing multi-hazards. Consequently, the present study introduces the production of a multi-hazard map for a settlement area by employing the Mamdani fuzzy inference algorithm.

3. Materials and Methods

Mamak District is a rapidly developing area located in the eastern part of Ankara, Turkey. Similar to many other large cities in Turkey, Ankara is affected by urban sprawl, and Mamak is one of the development centers for this sprawl. Approximately 640 thousand people live in the area. Since Mamak is on the route to Eastern Turkey, there is substantial transportation infrastructure. Furthermore, Bayindir Dam, which is currently used as a recreational area, is also located here. A part of Mamak District which is prone to both landslides and flooding was selected as the study area since it features continuous urban expansion and infrastructure (Figure 1). The area is ca. 30 km^2 and the minimum and maximum altitudes are 924 m and 1284 m, respectively.

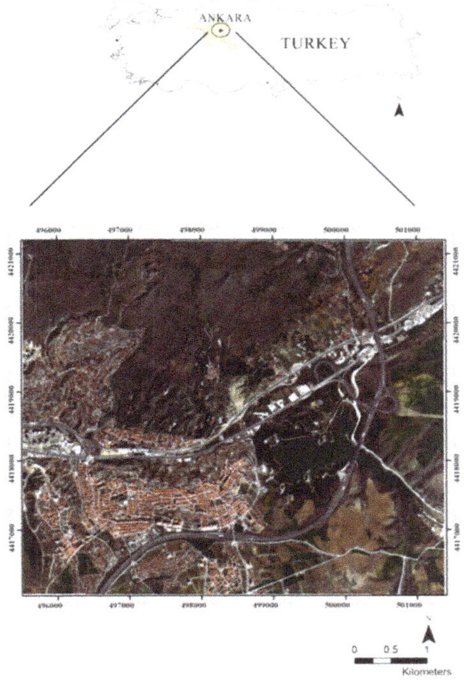

Figure 1. The location of the study area and an overview of the Sentinel-2 red–green–blue (RGB) image used in the study (upper left coordinates: 32°56′51.372″ E, 39°56′27.108″ N; lower right coordinates: 33°0′57.578″ E, 39°53′41.689″ N).

The landslide susceptibility map of the study area was produced by applying logistic regression (LR) to LULC data (sourced from Sentinel-2 imagery), the geomorphological features (sourced from DTM), and the lithology data [4]. Sentinel-2 images are distributed freely by the European Space Agency (ESA). It was found that actual land-use data can be produced from Sentinel-2 images, which are geometrically corrected (i.e., orthorectified, L2A) and obtained regularly over a large geographical extent by ESA [51]; they also provide multi-band (13 bands in total) data at spatial resolutions of 10 m, 20 m, and 60 m. The satellite constellation also has high transmission frequency [52] and, thus, is widely used in natural hazard assessments [53]. Sentinel-2 imagery can be easily employed by non-experts via the Sentinel Application Platform (SNAP) Tool from ESA. The LR is a rather simple method and can take non-numerical parameters such as different LULC types into account. The output probabilities were classified as low, moderate, and high. The flood susceptibility map of Ankara was produced in a previous study by Sozer et al. [19] for Ankara city and reclassified here into the same three classes (i.e., low, moderate, and high) to be used in the multi-hazard susceptibility assessment model. The study workflow is depicted in Figure 2. More details on the input data and methodology are provided in the sub-sections below.

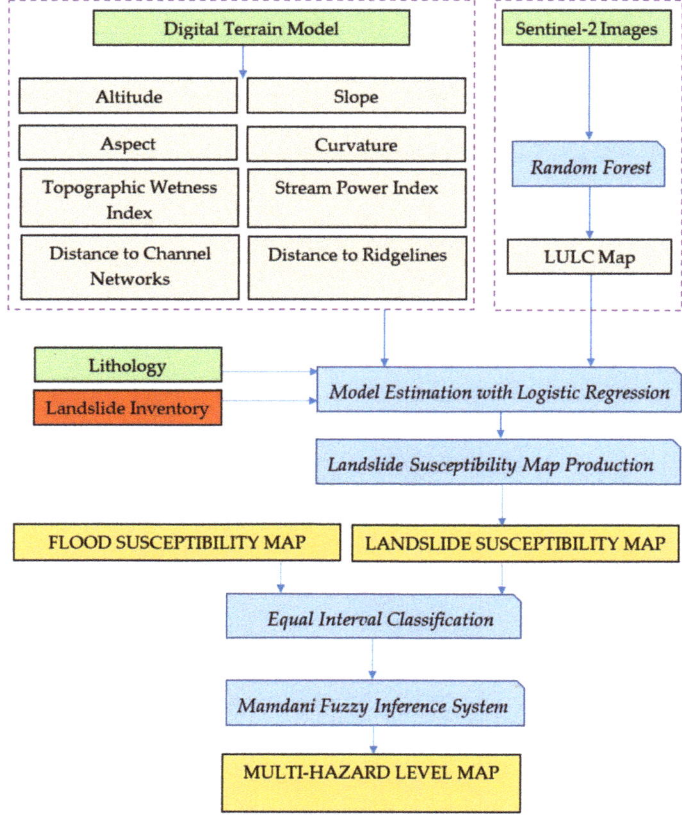

Figure 2. Overall workflow of the study.

3.1. Input Datasets for Landslide Susceptibility Map Production

A DTM with 5-m resolution was obtained from the General Directorate of Mapping (GDM), Turkey. The geomorphological parameters such as altitude, slope, general curvature, plan and profile curvatures, topographic wetness index (TWI), stream power index (SPI), distance to channel networks, and ridgelines were derived from the DTM. The Sentinel-2 satellite imagery from 23 March 2019 were employed for classifying the LULC using the RF method. The lithology data were digitized into vector form using the WebGIS portal data of the General Directorate of Mineral Research and Exploration (GDMRE/MTA), Turkey [54], and converted into raster data with 5-m grid spacing. The LULC map was also resampled to 5-m grid data to perform the LR technique. The data sources and the spatial resolutions are summarized in Table 1.

Table 1. Properties of input parameters. LULC—land use and land cover; DTM—digital terrain model; GDMRE/MTA—General Directorate of Mineral Research and Exploration.

Parameters	Source	Resolution
Geomorphological parameters	DTM	5 m
LULC	Sentinel-2 satellite imagery	10 m (resampled to 5 m)
Lithology	GDMRE/MTA	5 m
Landslide susceptibility map	Produced in the study	5 m

For training the LR method, boundary polygons of eight landslides covering ca. 2000 grid points were manually delineated by the expert using the DTM and the satellite images. The altitude range map and the landslide inventory represented by the red polygons are depicted in Figure 3. In order to utilize in the LR model estimation process, the vector landslide inventory map was rasterized with 5-m grid spacing. The landslides detected in the study area occurred in schists and volcanic units. These units are highly susceptible to weathering and landslides. The main characteristics of the landslides were circular, and the depth of failure surfaces was controlled by the thickness of weathered zones. In addition, the damage on the buildings in the study area was observed, and it is well known that the main cause of damage is the landslides in the area. However, it is impossible to draw the borders of the landslides due to urbanization on the slopes.

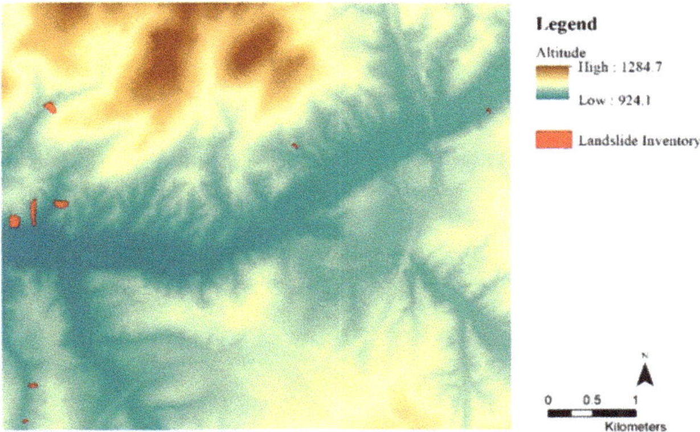

Figure 3. The elevation map and the manually delineated landslides (red polygons).

3.2. Geomorphological Characteristics of the Study Area

In order to understand the topography, some scalars such as primary and secondary derivatives of a DTM can be used. Primary features (e.g., slope, aspect, curvature) are computed from elevations, whereas secondary features (e.g. SPI, TWI) are obtained from the second derivatives of elevations [55]. These derivatives can be computed by using spatial analysis tools and software. Here, SAGA GIS from the SAGA User Group Association, Germany [56] and ArcGIS from ESRI Inc., Redlands, CA, USA [57] were used for this purpose. A statistical summary of the elevation data and the derivatives are given in Table 2. Similar statistics were derived for the landslide positive samples (~2000 grid points) and are provided in Table 3.

Table 2. Statistics of topographic attributes. SPI—stream power index; TWI—topographic wetness index.

Attribute Name	Minimum	Maximum	Mean	SD
Altitude (m)	924.1	1284.7	1032.2	62.8
Slope (°)	0.004	73.127	13.075	8.719
Aspect (°)	0	360	192.23	101.46
General curvature	−1.25957	1.09325	-9.73×10^{-5}	0.05887
Plan curvature	−0.09291	0.14917	4.56×10^{-4}	9.69×10^{-3}
Profile curvature	−0.16431	0.16666	-5.05×10^{-4}	0.01107
SPI	0	3,315,271.5	688.02	14,974.51
TWI	1.2776	22.1526	5.8651	2.1451
Distance to channel (m)	0.4	561.9	84.2	73.8
Distance to ridgeline (m)	0.0	229.9	33.0	26.5

Table 3. Statistics of topographic attributes in the landslides.

Attribute Name	Minimum	Maximum	Mean	SD
Altitude (m)	934.4	1050.4	986.5	30.7
Slope (°)	0.560	39.793	20.171	7.949
Aspect (°)	0.64	359.59	233.03	80.38
General curvature	−0.369	0.388	−0.0094	0.0943
Plan curvature	−0.0717	0.0493	−0.00418	0.0187
Profile curvature	−0.05139	0.04435	−0.00454	0.0139
SPI	0.362	9091.607	319.292	814.8529
TWI	2.2736	15.33	5.197	1.719
Distance to channel (m)	0.4	142.3	44.1	40.7
Distance to ridgeline (m)	0.0	50.4	15.7	10.9

Altitude shows the elevation measures of an area [58]. The slope gradient represents the variation in elevations [58] (Figure 4). Here, the slopes were employed to relate the topographical changes to landslide formation. The aspect was computed as the angle from the north to depict the direction of the slope. The aspect parameter was used to understand which slopes (north, south, etc.) would affect the landslides more [58]. The aspect values are also depicted in Figure 4.

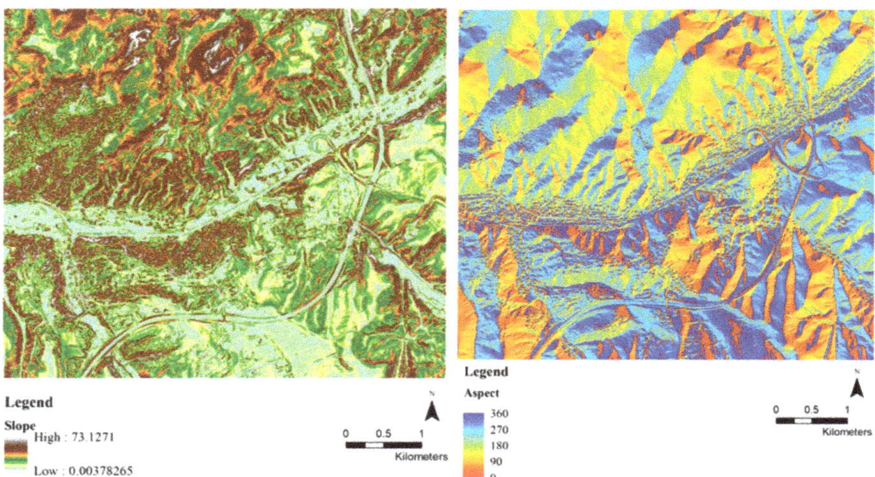

Figure 4. The slope gradient (**left**) and aspect (**right**) maps of the study area.

The curvature calculated from the DTM shows the changes in slope and aspect (Figure 5). The curvature parameter can be classified as plan and profile, and it needs to be analyzed separately. The planimetric component is based on the rate of slope variation along the contour lines, and the profile component is computed along the slope to determine the rate of slope gradient change [58]. Negative, positive, and zero curvatures reflect concave, convex, and flat surfaces, respectively [59]. The plan and profile curvatures of the study area are given in Figure 6.

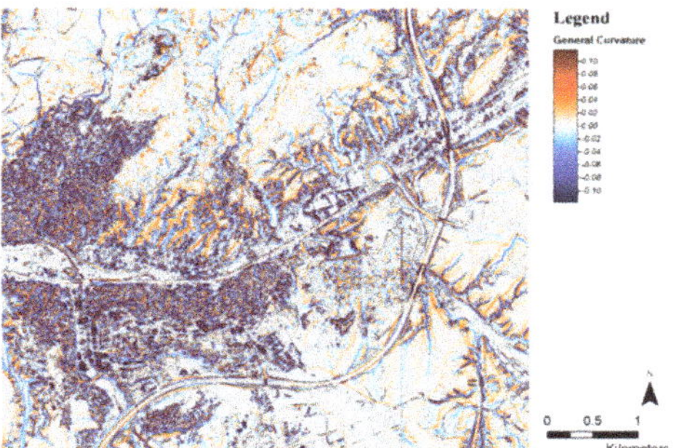

Figure 5. General curvature map of the study area.

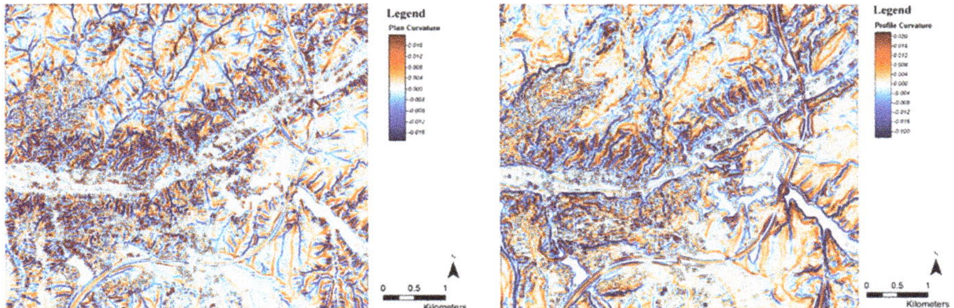

Figure 6. Plan (**left**) and profile (**right**) curvature maps of the study area.

The SPI is an indicator for erosive power of flowing water [60]. SPI is effective on landslides and denotes potential erosion energy [61]. The TWI shows the positions and size of the water-saturated regions [55] (Figure 7). The distances to channels were used for understanding the effect of the drainage network on landslides. The vertical distances to the channels were computed from the elevation data and, thus, the network was formed [62]. The landslide probability and the distances to the ridges were negatively correlated [59]. The ridges were computed using the DTM, and the distance values to both the ridges and the channel networks are provided in Figure 8.

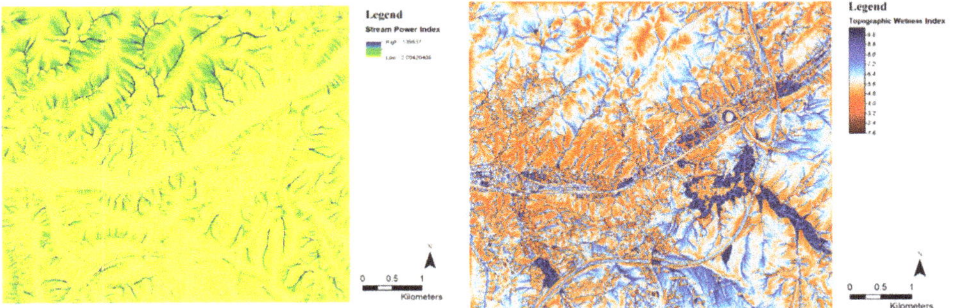

Figure 7. SPI (stream power index, on the left) and TWI (topographic wetness index, on the right) maps of the study area.

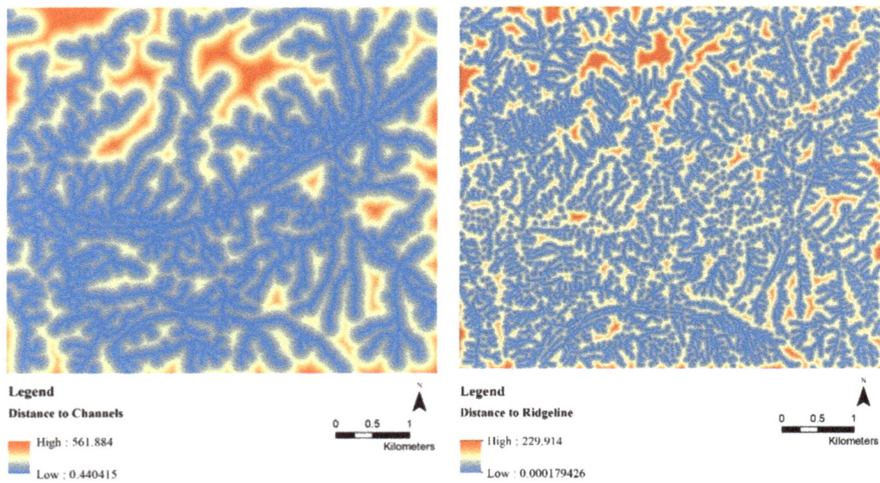

Figure 8. Distances to channels (**left**) and to ridgelines (**right**).

3.3. Land-Use and Land-Cover Extraction from Sentinel-2 Imagery

The use of up-to-date LULC data is essential for natural hazard assessments and disaster mitigation efforts. In this study, the LULC data were extracted from Sentinel-2 optical satellite imagery with 10-m spatial resolution. Seven LULC classes as shown in Figure 9 were extracted from red–green–blue (RGB) bands using SNAP software. The RF method was applied for the classification by collecting training data on the images. Using the RF method, classification is made with an ensemble of decision trees created by using training samples and variables [63]. In decision trees, the most rated pixels from all trees in the forest are classified. Because of the higher accuracy compared to other machine learning methods, it is widely used for image classification [64–67]. As seen in the study of Lim et al. [68] on Sentinel-2 images, the RF method can work with high accuracy in image classification. The RF classifier has two main parameters: the number of trees (T) and the number of variables (M) [69].

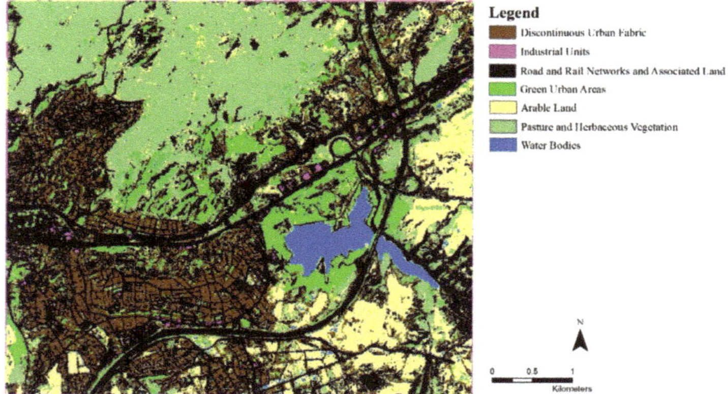

Figure 9. Land-use and land-cover map of the study area.

In this study, the classification process was completed by creating 10 trees using four bands, i.e., red, green, blue, and the gray-level co-occurrence matrix angular second moment (GLCM-ASM) parameter using 2077 training samples. Since it was difficult to separate the industrial units from the roads only with RGB information, the GLCM of the images was also computed and added to the classification as proposed by Stumpf and Kerle [70], and the GLCM-ASM parameter was found to be particularly useful in the present study. The distribution of the training samples to the classes is shown in Table 4. When the accuracy of classification was calculated by the cross-validation method, the correct prediction percentage was 93.73%, the classification precision was 92.01%, and the kappa value was 97.13% (Table 4). As can be seen in Table 4, the classification accuracies of all bands were improved by including the GLCM-ASM parameter, except for the industrial units, which remained high for both versions. In addition, it was visually verified that this parameter was especially useful to separate the road and industrial unit classes.

Table 4. Classification accuracy of the random forest method. ASM—angular second moment.

Land Use	Number of Training Samples	Classification Accuracy (Cross-Validation) (with ASM Band)	Classification Accuracy (Cross-Validation) (without ASM Band)
Discontinuous urban fabric	401	98.75%	94.01%
Industrial units	36	98.84%	98.94%
Road and rail networks and associated land	811	97.10%	94.4%
Green urban areas	146	99.71%	97.88%
Arable land	325	98.75%	96.72%
Pasture and herbaceous vegetation	125	98.65%	96.14%
Water bodies	233	99.81%	99.71%
Overall	2077	93.7259%	83.1081%
Kappa coefficient		97.13%	93.68%

3.4. Lithological Characteristics of the Study Area

In addition to the LULC and elevation data, the lithology type is extremely important for natural hazard assessments [71]. Lithology type and the structural differences generally affect the robustness and permeability of rocks and soils [72]. The lithology map was obtained from the geosciences portal (Yer Bilimleri Portali) of GDMRE/MTA. The lithology map is shown in Figure 10, and detailed descriptions are provided in Table 5. This vector map was preprocessed for conversion to raster data with 5-m grid spacing.

Table 5. Age and general descriptions of the lithologies in the study area [54].

Age	Description
Pliocene	Terrigenous clastics
Quaternary	Undifferentiated quaternary
Permian–Triassic	Clastics and carbonates
Upper Paleozoic Triassic	Schist, phyllite, marble, metabazite etc.
Lower–Middle Miocene	Non-graded volcanites

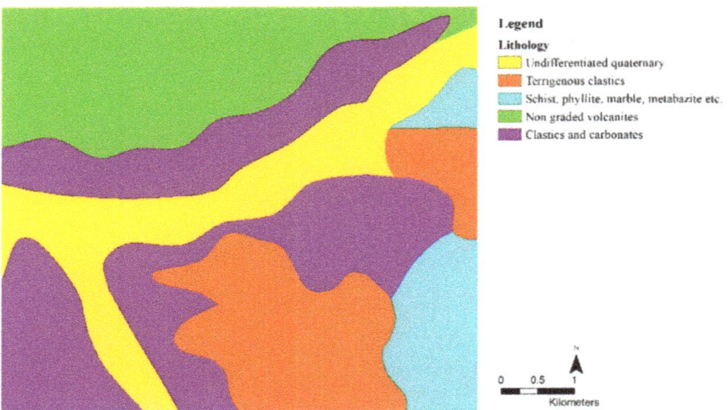

Figure 10. Lithology map of the study area [54].

3.5. Landslide Susceptibility Map Production with Logistic Regression Method

For the generation of landslide susceptibility maps, various mathematical and machine learning methodologies can be applied. Many studies on the assessment of landslide susceptibility using logistic regression were published in the literature (e.g., References [73–78]). In this study, multivariate LR was employed to derive the landslide susceptibility distribution of the area. LR is a statistical model, and it was used here to predict the potential landslide areas since it is fast and accurate for landslide susceptibility assessment purposes [9,59,79]. The LR is a supervised method and uses dependent (i.e., landslide conditioning factors) and independent (i.e., actual landslide inventory) variables. The dependent variable is a binary value which depicts the occurrence/non-occurrence of the event [62]. Independent variables were the 11 conditioning factors used as input data layers here (e.g., elevation data, slope, aspect, LULC, etc.). In the model estimation stage, the relationship between the variables was analyzed using the landslide positive samples (i.e., inventory data) and a number of randomly chosen non-landslide samples. Equations (1) and (2) were used for computing the logistic regression method.

$$Y_i = \beta_0 + \beta_1 X_i \tag{1}$$

$$P_i = (Y = 1|X_i) = 1/(1 + e^{\wedge}(-Y_i)) \tag{2}$$

where Y_i represents the dependent variables, x_i represents the independent variables, β_0 is a constant, β_i represents the *i*-th regression coefficient, and P is the probability of the existence of landslides [59]. In Vorpahl et al.'s [80] and Park et al.'s [81] studies, one of the most accurate results among the landslide susceptibility maps created with similar parameters used in this study was achieved using the LR method compared to other methods. After calculating the LR model parameters, the landslide susceptibility map was produced for the whole area. The ratio of the landslide positive and negative (non-landslide) samples was 1:2.

3.6. Flood Susceptibility Map of the Study Area

The flood susceptibility map of Ankara City was obtained from a previous study [19,20] (Figure 11) with the M-AHP method [21] using flow accumulation, slope, topographic altitude, distance to permanent river, distance to dry drainage, land cover, topographic wetness index, and lithology parameters. Each parameter was weighted with the M-AHP method according to expert opinion, and the highest score of each parameter was defined a priori. The output flood susceptibility map was reclassified by dividing the resulting probability values into three classes with equal intervals, and they were clipped for the study area (Figure 12). The original flood susceptibility map was divided into five equal classes by Sozer et al. [19]. However, in accordance with the purpose of the study, three classes of the flood susceptibility map were used in the present study. The histograms prepared for the five classes and three classes are given in Figure 13. The output susceptibility classes were categorized as low, moderate, and high, and they were used for the multi-hazard susceptibility assessment with the Mamdani fuzzy method.

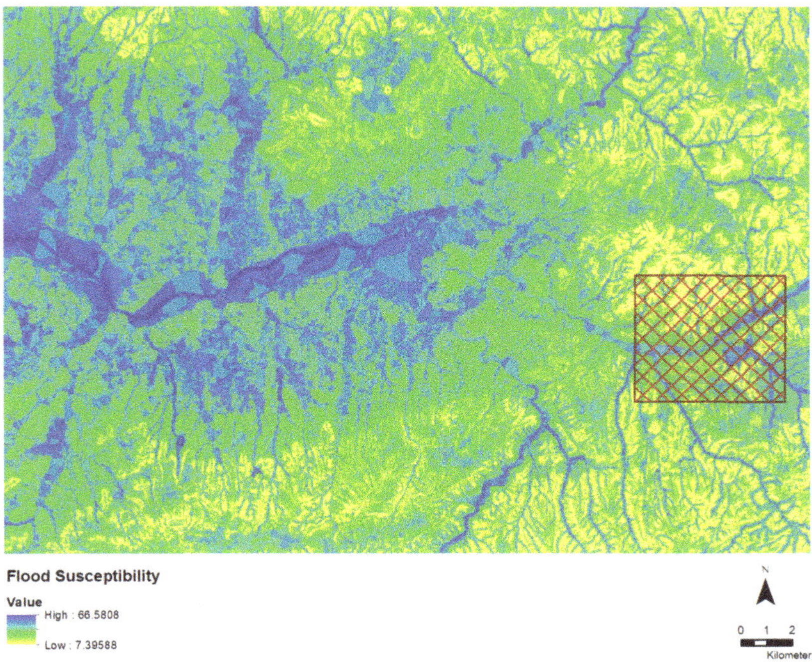

Figure 11. Flood susceptibility map produced by Sozer et al. [19] (the rectangular area to the east is the selected study area).

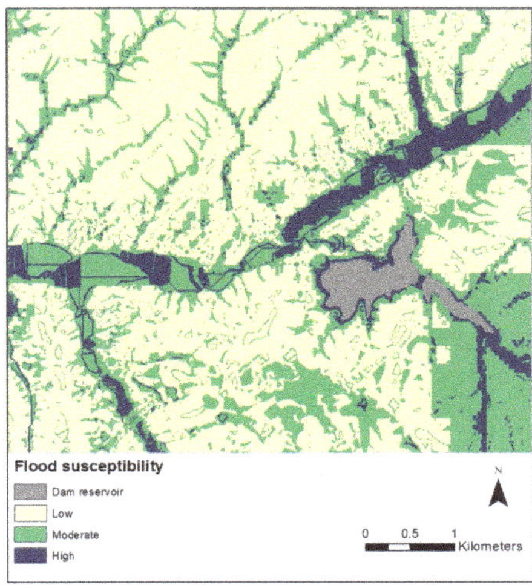

Figure 12. Flood susceptibility map of the study area (modified after Reference [19]).

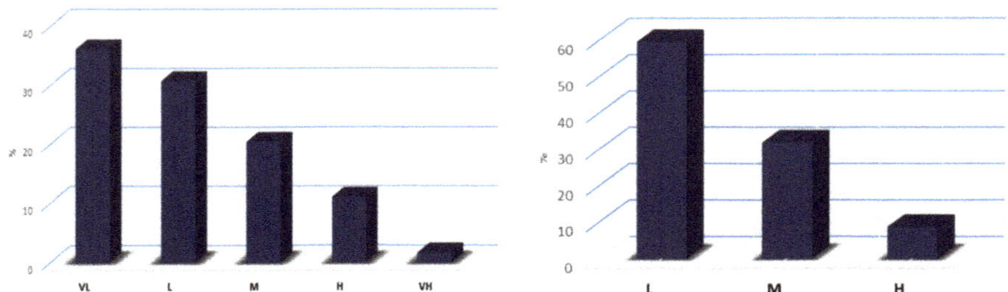

Figure 13. Histograms of flood susceptibility classes for five classes (**left**) and three classes (**right**).

3.7. Multi-Hazard Susceptibility Assesment with Mamdani Fuzzy Method

The multi-hazard susceptibility assessment map was derived with a combined assessment of flood and landslide susceptibility results using the Mamdani Fuzzy Method, which was first developed by Mamdani and Assilian [82]. This method is able to reduce uncertainties while solving complex problems using "if–then" rules. The stages of a Mamdani FIS are fuzzification, rule evaluation, aggregation, and defuzzification [82]. Fuzzy inference systems (FIS) produce a crisp output for supplied crisp inputs by using fuzzy set theory [83]. The general structure of a Mamdani FIS can be found in several books and publications (i.e., References [84–87]). Osna et al. [87] developed an integrated tool for construction of a Mamdani FIS for Netcad Software, Netcad, Ankara, Turkey. With the Mamdani fuzzy logic [88] operator in Netcad, landslide and flood susceptibilities could be evaluated together in the study area. The Mamdani fuzzy algorithm was previously used for landslide susceptibility mapping [86,87,89], but the present study is the first attempt at the combination of two susceptibility maps to obtain a multi-hazard susceptibility map.

The inputs of the Mamdani fuzzy model constructed in the study were landslide susceptibility and flood susceptibility maps, while the output was the multi-hazard susceptibility level (MHSL). Traditionally, a fuzzy model is built using expert knowledge in the form of linguistic rules. Three membership functions, i.e., low, moderate, and high, were defined for each input and output in the Mamdani FIS implemented here. The membership functions are shown in Figure 14, which also constitute the fuzzification stage of the system. In Figure 14, the vertical axes of the graphs denote the membership degree, and the horizontal axes represent the susceptibility levels, which range from 0–1 for landslide and 8–66 for flood. In the literature, many methods, such as intuition, rank ordering, angular fuzzy sets, genetic algorithms, inductive reasoning, soft partitioning, etc., exist for membership value assignment (e.g., References [90–92]). Although it is possible to select membership functions in a site-specific or target-oriented manner, or non-linearly, a generic approach was preferred here to prove the usability of the approach. In this study, the constructed fuzzy model employed two inputs and one output using three membership functions, and the fuzzification of crisp numbers and degree of membership of each crisp input were calculated at this stage. One of the fundamental features of a Mamdani FIS constitutes the linguistic if–then rules, namely, rule evaluation. In the present study, the if–then rules were generated by the expert (last author), which prevented an exhaustive data analysis process. The total number of linguistic rules generated by the expert was nine (Table 6). The final fuzzy output of the model was produced by aggregation of all local results from fuzzy rules triggered in the rule evaluation phase [87]. In the Mamdani FIS constructed here (Figure 15), the maximum operator was considered for aggregation, as suggested by Reference [87]. Finally, the center of gravity was used for defuzzification. Employing the Mamdani FIS constructed, the landslide and flood susceptibility maps were used as inputs, and the MHSL map was produced as presented in the section below.

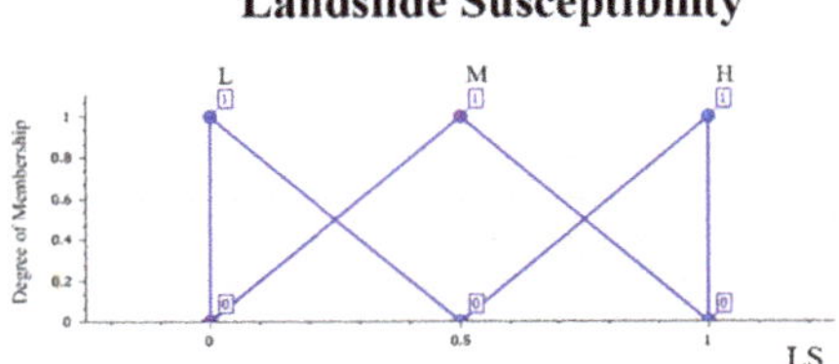

Figure 14. The membership functions of each input. The vertical axes in both graphs represent the degree of membership, while the horizontal axes reflect the susceptibility level range for landslide (**left**) and flood (**right**).

Table 6. *If–then* fuzzy rules used for the multi-hazard susceptibility level (MHSL) assessment in the study area.

Rule No.	Rule
1	If (landslide susceptibility is *high*) and (flood susceptibility is *high*), then (MHSL level is *high*),
2	If (landslide susceptibility is *high*) and (flood susceptibility is *moderate*), then (MHSL is *high*),
3	If (landslide susceptibility is *high*) and (flood susceptibility is *low*), then (MHSL is *high*)
4	If (landslide susceptibility is *moderate*) and (flood susceptibility is *high*), then (MHSL is *high*)
5	If (landslide susceptibility is *moderate*) and (flood susceptibility is *moderate*), then (MHSL is *high*)
6	If (landslide susceptibility is *moderate*) and (flood susceptibility is *low*), then (MHSL is *moderate*)
7	If (landslide susceptibility is *low*) and (flood susceptibility is *high*), then (MHSL is *high*)
8	If (landslide susceptibility is *low*) and (flood susceptibility is *moderate*), then (MHSL is *moderate*)
9	If (landslide susceptibility is *low*) and (flood susceptibility is *low*), then (MHSL is *low*)

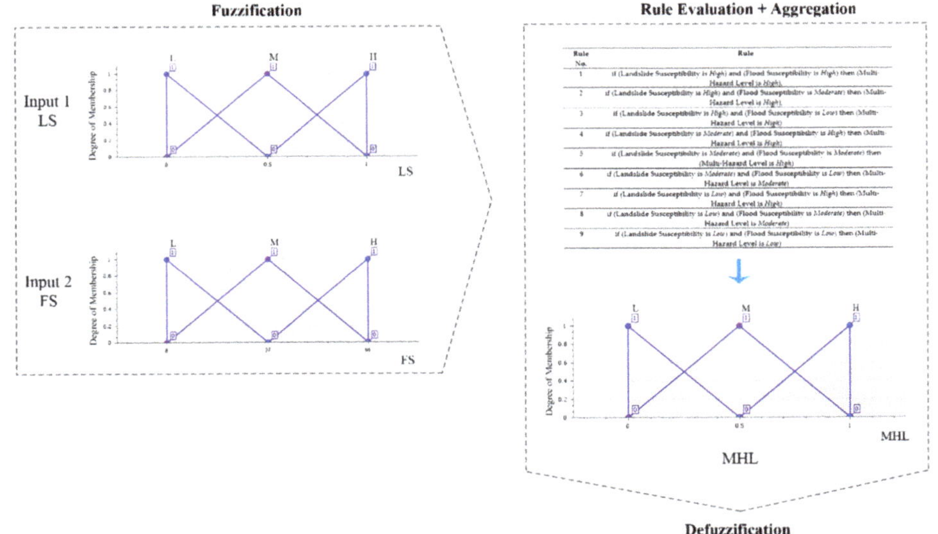

Figure 15. The general structure of the Mamdani fuzzy inference system (FIS) constructed.

4. Results and Discussion

4.1. The Landslide Susceptibility and MHSL Maps

The output landslide susceptibility map is shown in Figure 16. Although the landslide occurrence probability values ranged from 1%–99% (i.e., from 0–1, as shown in the landslide susceptibility membership graph in Figure 14), these values were reclassified into three categories (low, moderate, and high) by using equal interval classification for easy interpretation. These values were also used in the fuzzy assessment model in the next step. The map demonstrates the existing landslide hazard potential, especially in the western parts of the area. The field observations of the expert support the findings of the results obtained from the study.

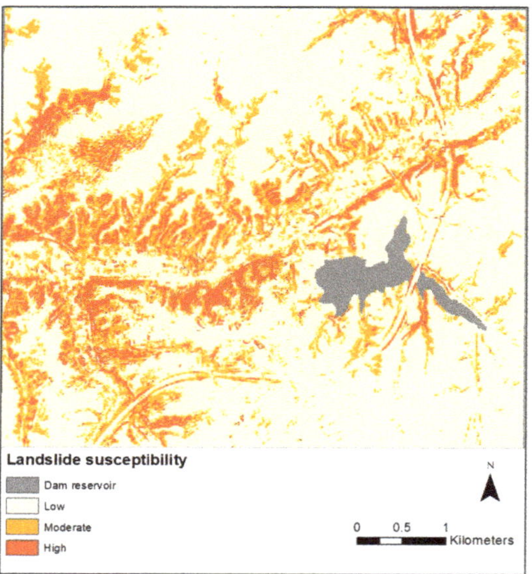

Figure 16. Landslide susceptibility map of the study area.

The accuracy of the output map was evaluated to understand the quality of the results. The factors that affect the accuracy are the data quality, the applied methods, the number of input parameters used in the process, and the approach for map production [72]. The ROC (receiver operating characteristic) curve, which is a measure of the capability of the current model in classification [93], was used to evaluate the accuracy (Figure 17). The figure shows that the areas with and without landslide susceptibility were classified with 96% accuracy.

In the present study, a plausible and practical methodology to combine different susceptibility maps was proposed. The Mamdani fuzzy algorithm was used and the MHSL map was obtained (Figure 18). The results show that some of the slopes and valleys have high multi-hazard potential. To minimize the losses caused by landslide and flood, the high multi-hazard susceptibility zones shown in Figure 18 must be investigated carefully, and necessary engineering measures must be provided at the construction stage.

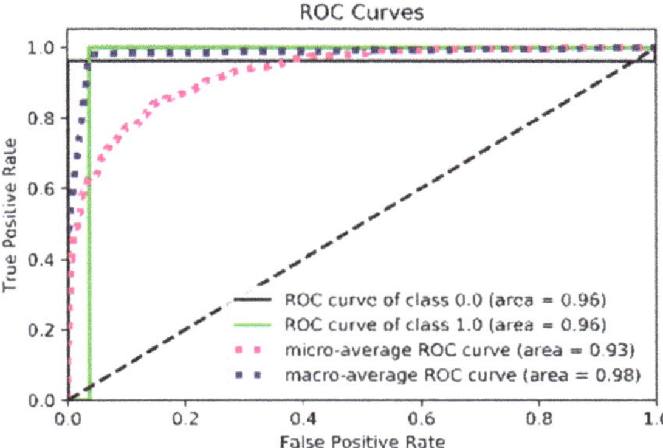

Figure 17. Receiver operating characteristic (ROC) curves of landslide susceptibility map.

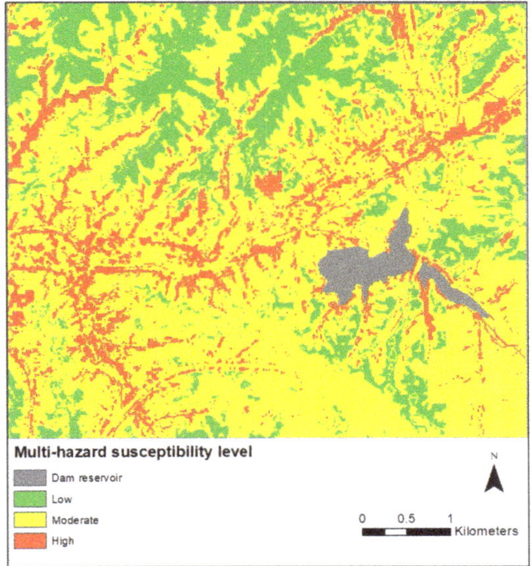

Figure 18. Multi-hazard susceptibility level map of the study area.

4.2. Discussion

The study area is subject to urban transformation projects due to unplanned settlements. The Mamak Urban Transformation Project was implemented in a part of the study area, which covers ca. 7.4 km^2. The project location is shown in Figure 19. It was divided into 11 stages and carried out by TOKI (Toplu Konut Idaresi Baskanligi), which is a state organization carrying out large-scale construction works for new houses, in the Ankara Metropolitan Municipality and Mamak Municipality. The main aim of the project was to transform the slums, i.e., unplanned settlement areas with insufficient facilities and infrastructure, and modernize these areas [94]. While the initial number of slums was 13,662, the number of slums destroyed as of 2019 was 8389 throughout the project. In total, 30,000 dwellings are planned to be constructed in the next phase of the project. Considering the natural hazard potential and related risks within the study area, the MHSL map could be used to analyze the vulnerability of future urban development and transformation plans.

Figure 19. A part of urban transformation project within the study area [91].

The DTM textured with the Sentinel-2 RGB image (Figure 20), the landslide susceptibility map (Figure 21), the flood susceptibility map (Figure 22), and the MHSL map (Figure 23) of the study area were visualized in 3D with the QT Modeler software from Applied Imagery, Chevy Chase, MD, USA [95] for interpretation of the results.

Figure 20. The DTM (digital terrain model) of the study area textured with the Sentinel-2 image.

Figure 21. The DTM of the study area textured with the landslide susceptibility map (output of logistic regression).

Figure 22. The DTM of the study area textured with the flood susceptibility map (modified into three classes after Sozer et al. [19]).

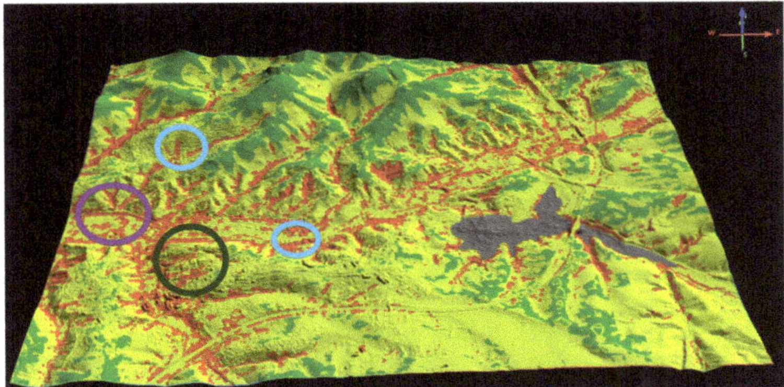

Figure 23. The DTM of the study area textured with the MHSL (multi-hazard susceptibility level) map. The circles denote important focal areas for city planning in northwest and south Mamak as mentioned above.

Figure 23 demonstrates the locations of the existing and the planned urban transformation sites. The existing project area (purple circle in Figure 23) includes areas with multi-hazard risk. When the risks are evaluated individually, there is low landslide risk in this area, but the excess of flood risk is also remarkable. In this respect, it can be considered to review the resilience of the project against natural hazard risks.

There are slum areas (blue circles in Figure 23) which represent potential urban transformation sites. Land-use decisions in these areas should be prepared elaborately. It is more appropriate to evaluate these areas at high susceptibility and utilize them as urban green areas by establishing agreements with the property owners. There are also newly constructed buildings (dark-green circle in Figure 23) in the west of the Mamak region. As can be seen, there are multiple areas with multi-hazard risk. The status of new buildings in these areas should be examined. The construction quality and the landslide resistance of these structures can be used as criteria to measure the accuracy of the previous partial urban transformation in this area.

With regard to the methodological approach employed here, the accuracy of the LR method was found sufficient for the purposes of the study. Although the number of the training samples in manually delineated landslide areas was low, using the 1:2 ratio for landslide/non-landslide samples worked efficiently. The produced MHSL map was representative for the purposes of the study, and it can be used as base map for urban planning and transformation purposes. The ensemble (i.e., LR) and expert-based (i.e., M-AHP) methods employed for the production of the individual hazard susceptibility maps were useful, and the Mamdani Fuzzy algorithm was able to handle the complexity of the problem.

5. Conclusions

In this study, a fuzzy model for integrated multi-hazard susceptibility assessment was developed. In addition, the usability of Sentinel-2 images in obtaining up-to-date LULC data in urban development areas for the production of landslide susceptibility maps was evaluated. The RF classification method was employed for producing LULC classes from Sentinel-2 RGB images, and the GLCM-ASM parameter was added to the classification to distinguish the industrial areas from roads. A high-resolution DTM and the lithology data were integrated into the landslide susceptibility map, and the LR method was applied for this purpose. An MHSL map was produced using the Mamdani fuzzy algorithm. The produced MHSL map can be used as essential data for urban development and transformation plans. Further analysis and planning can be carried out for this purpose.

The main difficulty encountered during the study was the preparation of a fully completed landslide inventory map due to urbanization. However, a combined methodology to obtain the MHSL map was described and applied successfully. The high multi-hazard susceptibility zones must be investigated carefully before construction; alternatively, if possible, these zones should be avoided with regard to construction purposes to minimize losses sourced from natural hazards. These maps are highly useful for planning stages, and, if these maps are considered during the planning stage, serious benefits can be obtained.

The methodology introduced in the present study for producing multi-hazard susceptibility represents a first in international literature. The use of an expert-based fuzzy inference system for the combination of two susceptibility maps portraying different natural hazards yielded very promising results. The study showed that the Mamdani type fuzzy inference system is a suitable approach for producing multi-hazard susceptibility mapping. the use of this approach for the combination of several multi-hazard susceptibility maps may provide new possibilities for suitable site selection efforts. However, there is no methodology available for the accurate assessment of a multi-hazard susceptibility map. For this reason, the study area selected was relatively small and from a well-known area. Consequently, the final output map was assessed with field observations.

As a final concluding remark, the fuzzy algorithm proposed for combining different natural hazards is a flexible and transparent modeling approach; hence, the model can be tuned or re-constructed easily when new information is obtained. As a future recommendation, some attempts at performance assessments of the multi-hazard susceptibility maps should be carried out, and some numerical indices for accuracy and performance assessments should be developed.

Author Contributions: Conceptualization and methodology, Candan Gokceoglu, Sultan Kocaman, and Tugce Yanar; investigation, validation, formal analysis, writing—original draft preparation, and visualization, Sultan Kocaman and Tugce Yanar; supervision, project administration, funding acquisition, and resources, Sultan Kocaman; data curation, Tugce Yanar; writing—review and editing, Candan Gokceoglu All authors have read and agreed to the published version of the manuscript.

Funding: This research is part of the M.Sc. thesis of the first author, and it was funded by the Hacettepe University Scientific Research Projects Coordination Unit (Grant FYL-2019-18017).

Acknowledgments: The authors sincerely thank the General Directorate of Mapping and the General Directorate of Mineral Research and Explorations for the provision of data, as well as Recep Can, Burhan Sozer, Orhan Firat, and Hakan A. Nefeslioglu from Hacettepe University for their continuous support. The authors would also like to thank the anonymous reviewers for their constructive comments and suggestions, which greatly helped to improve the manuscript.

Conflicts of Interest: The authors declare no conflicts of interest.

References

1. Kocaman, S.; Anbaroglu, B.; Gokceoglu, C.; Altan, O. A review on citizen science (CitSci) applications for disaster management. In *International Archives of the Photogrammetry, Remote Sensing & Spatial Information Sciences, Proceedings of the Geoinformation for Disaster Management (Gi4DM) Conference, Istanbul, Turkey, 18–21 March 2018*; ISPRS: Hannover, Germany, 2018; Volume XLII-3/W4, pp. 301–306. [CrossRef]
2. Mulero, D.L.; Nja, O.; Fernandez, C.L. Landslide Risk Management in the Urban Development of Sandnes (Norway). In *The International Archives of the Photogrammetry, Remote Sensing and Spatial Information Sciences, Proceedings of the Geoinformation for Disaster Management (Gi4DM) Conference, Istanbul, Turkey, 18–21 March 2018*; ISPRS: Hannover, Germany, 2018; Volume XLII-3/W4, pp. 327–334. [CrossRef]
3. Kocaman, S.; Gokceoglu, C. On the use of citsci and VGI in natural hazard assessment. In *International Archives of the Photogrammetry, Remote Sensing & Spatial Information Sciences, Proceedings of the ISPRS TC V Mid-Term Symposium Geospatial Technology—Pixel to People, Dehradun, India, 20–24 November 2018*; ISPRS: Hannover, Germany, 2018; Volume XLII-5, pp. 69–73. [CrossRef]
4. Yanar, T.; Kocaman, S.; Gokceoglu, C. On The use of sentinel-2 images and high resolution DTM For landslide susceptibility mapping in a developing urban settlement (Mamak, Ankara, Turkey). In *International Archives of the Photogrammetry, Remote Sensing & Spatial Information Sciences, Proceedings of the Geoinformation for Disaster Management (Gi4DM), Prague, Czech Republic, 3–6 September 2019*; ISPRS: Hannover, Germany, 2019; Volume XLII-3/W8, pp. 469–476. [CrossRef]
5. AFAD. Afet ve Acil Durum Yönetimi Başkanlığı. Available online: https://www.afad.gov.tr/ (accessed on 1 December 2019).
6. Nefeslioglu, H.A.; San, B.T.; Gokceoglu, C.; Duman, T.Y. An assessment on the use of Terra ASTER L3A data in landslide susceptibility mapping. *Int. J. Appl. Earth Obs. Geoinf.* **2012**, *14*, 40–60. [CrossRef]
7. Pham, B.T.; Shirzadi, A.; Tien Bui, D.; Prakash, I.; Dholakia, M. A hybrid machine learning ensemble approach based on a radial basis function neural network and rotation forest for landslide susceptibility modeling: A case study in the Himalayan area, India. *Int. J. Sediment Res.* **2018**, *33*, 157–170. [CrossRef]
8. Gorum, T.; Gonencgil, B.; Gokceoglu, C.; Nefeslioglu, H.A. Implementation of reconstructed geomorphologic units in landslide susceptibility mapping: The Melen Gorge (NW Turkey). *Nat. Hazards* **2008**, *46*, 323–351. [CrossRef]
9. Reichenbach, P.; Rossi, M.; Malamud, B.D.; Mihir, M.; Guzzetti, F. A review of statistically-based landslide susceptibility models. *Earth Sci. Rev.* **2018**, *180*, 60–91. [CrossRef]
10. Sevgen, E.; Kocaman, S.; Nefeslioglu, H.A.; Gokceoglu, C. A novel performance assessment approach using photogrammetric techniques for landslide susceptibility mapping with logistic regression, ANN and random forest. *Sensors* **2019**, *19*, 3940. [CrossRef] [PubMed]
11. Kocaman, S.; Gokceoglu, C. A CitSci app for landslide data collection. *Landslides* **2019**, *16*, 611–615. [CrossRef]
12. Chen, L.; Guo, Z.; Yin, K.; Shrestha, D.P.; Jin, S. The influence of land use and land cover change on landslide susceptibility: A case study in Zhushan Town, Xuanen County (Hubei, China). *Nat. Hazards Earth Syst. Sci. Discussions* **2019**. [CrossRef]
13. Adhikari, P.; Hong, Y.; Douglas, K.R.; Kirschbaum, D.B.; Gourley, J.; Adler, R.; Brakenridge, G.R. A digitized global flood inventory (1998–2008): Compilation and preliminary results. *Nat. Hazards* **2010**, *55*, 405–422. [CrossRef]
14. CRED. Natural Disasters 2017. Available online: https://cred.be/sites/default/files/adsr_2017.pdf (accessed on 1 December 2019).
15. Gökçe, O.; Özden, Ş.; Demir, A. *Türkiye'de Afetlerin Mekansal ve İstatistiksel Dağılımı Afet Bilgileri Envanteri*; Bayındırlık ve İskân Bakanlığı: Ankara, Turkey, 2008.
16. Perucca, L.P.; Angilieri, Y.E. Morphometric characterization of del Molle Basin applied to the evaluation of flash floods hazard, Iglesia Department, San Juan, Argentina. *Quat. Int.* **2011**, *233*, 81–86. [CrossRef]
17. Scheuer, S.; Haase, D.; Meyer, V. Towards a flood risk assessment ontology—Knowledge integration into a multi-criteria risk assessment approach. *Comput. Environ. Urban Syst.* **2013**, *37*, 82–94. [CrossRef]
18. Cunha, N.; Magalh, M.; Domingos, T.; Abreu, M.; Küpfer, C. The land morphology approach to flood risk mapping: An application to Portugal. *J. Environ. Manag.* **2017**, *193*, 172–187. [CrossRef] [PubMed]

19. Sozer, B.; Kocaman, S.; Nefeslioglu, H.A.; Firat, O.; Gokceoglu, C. Preliminary investigations on flood susceptibility mapping in Ankara (Turkey) using modified analytical hierarchy process (M-AHP). *Int. Arch. Photogramm. Remote Sens. Spat. Inf. Sci.* **2018**, *42*, 5. [CrossRef]
20. Sözer, B.; Kocaman, S.; Nefeslioğlu, H.A.; Fırat, O.; Gökçeoğlu, C. Değiştirilmiş AHP (M-AHP) yöntemi kullanilarak ankara için taşkin duyarlilik haritasi üretimi. *Harit. Derg.* **2019**, *162*, 12–24.
21. Nefeslioglu, H.A.; Sezer, E.A.; Gokceoglu, C.; Ayas, Z. A Modified Analytical Hierarchy Process (M-AHP) approach for decision support systems in natural hazard assessments. *Comput. Geosci.* **2013**, *59*, 1–8. [CrossRef]
22. Basheer Ahammed, K.K.; Pandey, A. Geoinformatics based assessment of coastal multi-hazard vulnerability along the East Coast of India. *Spat. Inf. Res.* **2019**, *27*, 295–307. [CrossRef]
23. Bathrellos, G.; Skilodimou, H.; Chousianitis, K.; Youssef, A.; Pradhan, B. Suitability estimation for urban development using multi-hazard assessment map. *Sci. Total Environ.* **2017**, 119–134. [CrossRef]
24. Skilodimou, H.; Bathrellos, G.; Chousianitis, K.; Youssef, A.; Pradhan, B. Multi-hazard assessment modeling via multi-criteria analysis and GIS: A case study. *Environ. Earth Sci.* **2019**, 78. [CrossRef]
25. Bani-Mustafa, T.; Zeng, Z.; Zio, E.; Vasseur, D. A framework for multi-hazards risk aggregation considering risk model maturity levels. In Proceedings of the IEEE 2nd International Conference on System Reliability and Safety (ICSRS), Milan, Italy, 20–22 December 2017; pp. 429–433. [CrossRef]
26. Furlan, E.; Torresan, S.; Critto, A.; Marcomini, A. Spatially explicit risk approach for multi-hazard assessment and management in marine environment: The case study of the Adriatic Sea. *Sci. Total Environ.* **2017**, *618*, 1008–1023. [CrossRef]
27. Mukhopadhyay, A.; Hazra, S.; Mitra, D.; Hutton, C.; Chanda, A.; Mukherjee, S. Characterizing the multi-risk with respect to plausible natural hazards in the Balasore coast, Odisha, India: A multi-criteria analysis (MCA) appraisal. *Nat. Hazards* **2016**, *80*, 1495–1513. [CrossRef]
28. Chen, L.; van Westen, C.J.; Hussin, H.; Ciurean, R.L.; Turkington, T.; Chavarro-Rincon, D.; Shrestha, D.P. Integrating expert opinion with modelling for quantitative multi-hazard risk assessment in the Eastern Italian Alps. *Geomorphology* **2016**, *273*, 150–167. [CrossRef]
29. Ehlen, M.; Vargas, V. Multi-hazard, multi-infrastructure, economic scenario analysis. *Environ. Syst. Decis.* **2013**, *33*, 60–75. [CrossRef]
30. Zhou, Y.; Liu, Y.; Wu, W.; Ning, L. Integrated risk assessment of multi-hazards in China. *Nat. Hazards* **2015**, *78*, 257–280. [CrossRef]
31. Tian, C.; Yiping, F.; Yang, L.; Zhang, C. Spatial-temporal analysis of community resilience to multi-hazards in the Anning River basin, Southwest China. *Int. J. Disaster Risk Reduct.* **2019**, *39*, 101144. [CrossRef]
32. Gill, J.; Malamud, B. Anthropogenic processes, natural hazards, and interactions in a multi-hazard framework. *Earth Sci. Rev.* **2017**, *166*, 246–269. [CrossRef]
33. Gallina, V.; Torresan, S.; Critto, A.; Sperotto, A.; Glade, T.; Marcomini, A. A review of multi-risk methodologies for natural hazards: Consequences and challenges for a climate change impact assessment. *J. Environ. Manag.* **2015**, *168*, 123–132. [CrossRef] [PubMed]
34. Barrantes, G. Multi-hazard model for developing countries. *Nat. Hazards* **2018**, *92*, 1081–1095. [CrossRef]
35. Liu, B.; Siu, Y.L.; Mitchell, G. A quantitative model for estimating risk from multiple interacting natural hazards: An application to northeast Zhejiang, China. *Stoch. Environ. Res. Risk Assess.* **2016**, *31*, 1319–1340. [CrossRef]
36. Bernal, G.; Salgado-Gálvez, M.; Zuloaga Romero, D.; Tristancho, J.; González, D.; Cardona, O. Integration of probabilistic and multi-hazard risk assessment within urban development planning and emergency preparedness and response: Application to Manizales, Colombia. *Int. J. Disaster Risk Sci.* **2017**, *8*, 270–283. [CrossRef]
37. Micu, M.; Jurchescu, M.; Micu, D.; Zarea, R.; Zumpano, V.; Bălteanu, D. A morphogenetic insight into a multi-hazard analysis: Bâsca Mare landslide dam. *Landslides* **2014**, *11*, 1131–1139. [CrossRef]
38. Tilloy, A.; Malamud, B.; Winter, H.; Joly-Laugel, A. A review of quantification methodologies for multi-hazard interrelationships. *Earth Sci. Rev.* **2019**, *196*, 102881. [CrossRef]
39. Chang, S.; Yip, J.; Tse, W. Effects of urban development on future multi-hazard risk: The case of Vancouver, Canada. *Nat. Hazards* **2018**, *98*, 251–265. [CrossRef]
40. Mafi Gholami, D.; Zenner, E.; Jaafari, A.; Riyahi Bakhtyari, H.R.; Tien Bui, D. Multi-hazards vulnerability assessment of southern coasts of Iran. *J. Environ. Manag.* **2019**, *252*, 109628. [CrossRef] [PubMed]

41. Fell, R.; Jordi, C.; Christophe, B.; Leonardo, C.; Eric, L.; William, Z.S. Guidelines for landslide susceptibility, hazard and risk zoning for land use planning. *Eng. Geol.* **2008**, *102*, 85–98. [CrossRef]
42. Jacobs, L.; Maes, J.; Mertens, K.; Sekajugo, J.; Thiery, W.; Lipzig, N.; Poesen, J.; Kervyn, M.; Dewitte, O. Reconstruction of a flash flood event through a multi-hazard approach: Focus on the Rwenzori Mountains, Uganda. *Nat. Hazards* **2016**, *84*, 851–876. [CrossRef]
43. Kappes, M.; Papathoma-Köhle, M.; Keiler, M. Assessing physical vulnerability for multi-hazards using an indicator-based methodology. *Appl. Geogr.* **2012**, *32*, 577–590. [CrossRef]
44. Omidvar, B.; Karimi, H. Multi-hazard failure probability analysis of gas pipelines for earthquake shaking, ground failure and fire following earthquake. *Nat. Hazards* **2016**, *82*, 703–720. [CrossRef]
45. Liu, K.; Wang, M.; Cao, Y.; Zhu, W.; Yang, G. Susceptibility of existing and planned Chinese railway system subjected to rainfall-induced multi-hazards. *Transp. Res. Part A Policy Pract.* **2018**, *117*, 214–226. [CrossRef]
46. Mirzaei, G.; Soltani, A.; Soltani, M.; Darabi, M. An integrated data-mining and multi-criteria decision-making approach for hazard-based object ranking with a focus on landslides and floods. *Environ. Earth Sci.* **2018**, *77*, 581. [CrossRef]
47. Sheikh, V.; Kornejady, A.; Ownegh, M. Application of the coupled TOPSIS-mahalanobis distance for multi-hazard-based management of the target districts of the Golestan Province, Iran. *Nat. Hazards* **2019**, *96*, 1335–1365. [CrossRef]
48. Pourghasemi, H.R.; Gayen, A.; Panahi, M.; Rezaie, F.; Blaschke, T. Multi-hazard probability assessment and mapping in Iran. *Sci. Total Environ.* **2019**, *692*, 556–571. [CrossRef]
49. Araya-Munoz, D.; Metzger, M.; Stuart, N.; Wilson, A.; Carvajal, D. A spatial fuzzy logic approach to urban multi-hazard impact assessment in Concepción, Chile. *Sci. Total Environ.* **2017**, *576*, 508–519. [CrossRef] [PubMed]
50. Kappes, M.S.; Keiler, M.; von Elverfeldt, K.; Glade, T. Challenges of analyzing multi-hazard risk: A review. *Nat. Hazards* **2012**, *64*, 1925–1958. [CrossRef]
51. ESA. Sentinel-2. Available online: https://sentinel.esa.int/web/sentinel/missions/sentinel-2 (accessed on 27 January 2019).
52. ESA. Sentinel-2 Coverage Map. Available online: https://sentinel.esa.int/web/sentinel/user-guides/sentinel-2-msi/revisit-coverage (accessed on 27 January 2019).
53. Poursanidis, D.; Chrysoulakis, N. Remote Sensing, Natural Hazards and the Contribution of ESA Sentinels Missions. *Remote Sens. Appl. Soc. Environ.* **2017**, *6*, 25–38. [CrossRef]
54. Akbaş, B.; Akdeniz, N.; Aksay, A.; Altun, İ.; Balcı, V.; Bilginer, E.; Bilgiç, T.; Duru, M.; Ercan, T.; Gedik, İ.; et al. *Diğerleri Türkiye Jeoloji Haritasi*; Maden Tetkik ve Arama Genel Müdürlüğü Yayını: Ankara, Turkey, 2002; Available online: http://yerbilimleri.mta.gov.tr/ (accessed on 20 July 2018).
55. Moore, I.D.; Grayson, R.B.; Ladson, A.R. Digital terrain modelling: A review of hydrological, geomorphological, and biological applications. *Hydrol. Process.* **1991**, *5*, 3–30. [CrossRef]
56. SAGA GIS. System for Automated Geoscientific Analyses. Available online: www.saga-gis.org (accessed on 20 July 2019).
57. ESRI. Environmental Systems Research Institute, Inc. Available online: https://www.esri.com/ (accessed on 20 July 2018).
58. Wilson, J.P.; Gallant, J.C. *Terrain Analysis: Principles and Applications*; John Wiley and Sons, Inc.: Ottawa, ON, Canada, 2000.
59. Budimir, M.; Atkinson, P.; Lewis, H. Seismically induced landslide hazard and exposure modelling in Southern California based on the 1994 Northridge, California earthquake event. *Landslides* **2014**, *12*, 1–16. [CrossRef]
60. Zakerinejad, R.; Maerker, M. An integrated assessment of soil erosion dynamics with special emphasis on gully erosion in the Mazayjan basin, southwestern Iran. *Nat. Hazards* **2015**, *79*, 25–50. [CrossRef]
61. Kakembo, V.; Xanga, W.W.; Rowntree, K. Topographic thresholds in gully development on the hillslopes of communal areas in Ngqushwa local municipality, Eastern Cape, South Africa. *Geomorphology* **2009**, *110*, 188–194. [CrossRef]
62. Samia, J.; Temme, A.; Bregt, A.; Wallinga, J.; Guzzetti, F.; Ardizzone, F.; Rossi, M. Characterization and quantification of path dependency in landslide susceptibility. *Geomorphology* **2017**, *292*, 16–24. [CrossRef]
63. Breiman, L. Random forests. *Mach. Learn.* **2001**, *45*, 5–32. [CrossRef]

64. Belgiu, M.; Drăguţ, L. Random forest in remote sensing: A review of applications and future directions. *ISPRS J. Photogramm. Remote Sens.* **2016**, *114*, 24–31. [CrossRef]

65. Stefanski, J.; Mack, B.; Waske, O. Optimization of object-based image analysis with random forests for land cover mapping. *IEEE J. Sel. Top. Appl. Earth Obs. Remote Sens.* **2013**, *6*, 2492–2504. [CrossRef]

66. Yoo, C.; Han, D.; Im, J.; Bechtel, B. Comparison between convolutional neural networks and random forest for local climate zone classification in mega urban areas using Landsat images. *ISPRS J. Photogramm. Remote Sens.* **2019**, *157*, 155–170. [CrossRef]

67. Ma, L.; Li, M.; Ma, X.; Cheng, L.; Du, P.; Liu, Y. A review of supervised object-based land-cover image classification. *ISPRS J. Photogramm. Remote Sens.* **2017**, *130*, 277–293. [CrossRef]

68. Lim, J.; Kim, K.; Jin, R. Tree species classification using hyperion and sentinel-2 data with machine learning in south Korea and China. *Int. J. Geo Inf.* **2019**, *8*, 150. [CrossRef]

69. Guo, L.; Chehata, N.; Mallet, C.; Boukir, S. Relevance of airborne lidar and multispectral image data for urban scene classification using Random Forests. *ISPRS J. Photogramm. Remote Sens.* **2011**, *66*, 56–66. [CrossRef]

70. Stumpf, A.; Kerle, N. Object-oriented mapping of landslides using random forests. *Remote Sens. Environ.* **2011**, *115*, 2564–2577. [CrossRef]

71. Pourghasemi, H.R.; Kerle, N. Random forests and evidential belief function-based landslide susceptibility assessment in Western Mazandaran Province, Iran. *Environ. Earth Sci.* **2016**, *75*. [CrossRef]

72. Ayalew, L.; Yamagishi, H. The application of GIS-based logistic regression for landslide susceptibility mapping in the Kakuda-Yahiko Mountains, Central Japan. *Geomorphology* **2005**, *65*, 15–31. [CrossRef]

73. Lee, S. Application of logistic regression model and its validation for landslide susceptibility mapping using GIS and remote sensing data. *Int. J. Remote Sens.* **2005**, *26*, 1477–1491. [CrossRef]

74. Duman, T.Y.; Can, T.; Gokceoglu, C.; Nefeslioglu, H.A.; Sonmez, H. Application of logistic regression for landslide susceptibility zoning of Cekmece Area, Istanbul, Turkey. *Environ. Geol.* **2006**, *51*, 241–256. [CrossRef]

75. Wang, H.B.; Sassa, K. Comparative evaluation of landslide susceptibility in Minamata area, Japan. *Environ. Geol.* **2005**, *47*, 956–966. [CrossRef]

76. Bai, S.-B.; Wang, J.; Lü, G.-N.; Zhou, P.-G.; Hou, S.-S.; Xu, S.-N. GIS-based logistic regression for landslide susceptibility mapping of the Zhongxian segment in the Three Gorges area, China. *Geomorphology* **2010**, *115*, 23–31. [CrossRef]

77. Das, I.; Sahoo, S.; Van Westen, C.; Stein, A.; Hack, R. Landslide susceptibility assessment using logistic regression and its comparison with a rock mass classification system, along a road section in the northern Himalayas (India). *Geomorphology* **2010**, *114*, 627–637. [CrossRef]

78. Nefeslioglu, H.A.; Gokceoglu, C. Probabilistic risk assessment in medium scale for rainfall-induced earthflows: Catakli catchment area (Cayeli, Rize, Turkey). *Math. Probl. Eng.* **2011**. [CrossRef]

79. Lee, S.; Pradhan, B. Landslide hazard mapping at Selangor, Malaysia using frequency ratio and logistic regression models. *Landslides* **2006**, *4*, 33–41. [CrossRef]

80. Vorpahl, P.; Elsenbeer, H.; Märker, M.; Schröder, B. How can statistical models help to determine driving factors of landslides? *Ecol. Model.* **2012**, *239*, 27–39. [CrossRef]

81. Park, S.; Choi, C.; Kim, B.; Kim, J. Landslide susceptibility mapping using frequency ratio, analytic hierarchy process, logistic regression, and artificial neural network methods at the Inje area, Korea. *Environ. Earth Sci.* **2012**, *68*, 1443–1464. [CrossRef]

82. Mamdani, E.H.; Assilian, S. Experiment in linguistic synthesis with a fuzzy logic controller. *Int. J. Man Mach. Stud.* **1975**, *7*, 1–13. [CrossRef]

83. Zadeh, L.A. Fuzzy sets. *Inf. Control* **1965**, *8*, 338–353. [CrossRef]

84. Grima, M.A. *Neuro-Fuzzy Modeling in Engineering Geology*; A.A. Balkema: Rotterdam, The Netherlands, 2000; p. 244. Available online: https://books.google.com.tr/books/about/Neuro_fuzzy_Modeling_in_Engineering_Geol.html?id=PJ6VQgAACAAJ&source=kp_cover&redir_esc=y (accessed on 17 February 2020).

85. Gokceoglu, C.; Zorlu, K. A fuzzy model to predict the uniaxial compressive strength and the modulus of elasticity of a problematic rock. *Eng. Appl. Artif. Intell.* **2004**, *17*, 61–72. [CrossRef]

86. Akgun, A.; Sezer, E.A.; Nefeslioglu, H.A.; Gokceoglu, C.; Pradhan, B. An easy-to-use MATLAB program (MamLand) for the assessment of landslide susceptibility using a Mamdani fuzzy algorithm. *Comput. Geosci.* **2012**, *38*, 23–34. [CrossRef]

87. Osna, T.; Sezer, E.A.; Akgun, A. GeoFIS: An integrated tool for the assessment of landslide susceptibility. *Comput. Geosci.* **2014**, *66*, 20–30. [CrossRef]
88. Sezer, E.A.; Nefeslioglu, H.A.; Osna, T. An expert-based landslide susceptibility mapping (LSM) module developed for Netcad Architect Software. *Comput. Geosci.* **2017**, *98*, 26–37. [CrossRef]
89. Ozer, B.C.; Mutlu, B.; Nefeslioglu, H.; Sezer, E.; Rouai, M.; Dekayir, A.; Gokceoglu, C. On the use of hierarchical fuzzy inference systems (HFIS) in expert-based landslide susceptibility mapping: The central part of the Rif Mountains (Morocco). *Bull. Eng. Geol. Environ.* **2020**, *79*, 551–568. [CrossRef]
90. Zadeh, L.A. A Rationale for fuzzy control. *J. Dyn. Syst. Meas. Control Trans. ASME* **1972**, *94*, 3–4. [CrossRef]
91. Hadipriono, F.; Sun, K. Angular fuzzy set models for linguistic values. *Civ. Eng. Syst.* **1990**, *7*, 148–156. [CrossRef]
92. Karr, C.L.; Gentry, E.J. Fuzzy control of pH using genetic algorithms. *IEEE Trans. Fuzzy Syst.* **1993**, *1*, 46–53. [CrossRef]
93. Perlich, C.; Provost, F.; Simonoff, J. Tree induction vs. logistic regression: A learning-curve analysis. *J. Mach. Learn. Res.* **2003**, *4*, 211–255. [CrossRef]
94. Belediyesi, M. Mamak Belediyesi 2018 Yılı İdare Faaliyet Raporu. Available online: https://www.mamak.bel.tr/wp-content/uploads/2017/03/2018-faaliyet.pdf (accessed on 20 July 2019).
95. QT Modeler. Available online: http://appliedimagery.com/ (accessed on 29 December 2019).

 International Journal of *Geo-Information*

Article

Mapping Impact of Tidal Flooding on Solar Salt Farming in Northern Java using a Hydrodynamic Model

Anang Widhi Nirwansyah [1,2,*] and Boris Braun [2]

[1] Department of Geography, Universitas Muhammadiyah Purwokerto, Banyumas 53182, Indonesia
[2] Institute of Geography, University of Cologne, 50923 Cologne, Germany; boris.braun@uni-koeln.de
* Correspondence: anangwidi@ump.ac.id

Received: 19 August 2019; Accepted: 10 October 2019; Published: 12 October 2019

Abstract: The number of tidal flood events has been increasing in Indonesia in the last decade, especially along the north coast of Java. Hydrodynamic models in combination with Geographic Information System applications are used to assess the impact of high tide events upon the salt production in Cirebon, West Java. Two major flood events in June 2016 and May 2018 were selected for the simulation within inputs of tidal height records, national seamless digital elevation dataset of Indonesia (DEMNAS), Indonesian gridded national bathymetry (BATNAS), and wind data from OGIMET. We used a finite method on MIKE 21 to determine peak water levels, and validation for the velocity component using TPXO9 and Tidal Model Driver (TMD). The benchmark of the inundation is taken from the maximum water level of the simulation. This study utilized ArcGIS for the spatial analysis of tidal flood distribution upon solar salt production area, particularly where the tides are dominated by local factors. The results indicated that during the peak events in June 2016 and May 2018, about 83% to 84% of salt ponds were being inundated, respectively. The accurate identification of flooded areas also provided valuable information for tidal flood assessment of marginal agriculture in data-scarce region.

Keywords: mapping impact; tidal flood; hydrodynamic model; solar salt farming

---

1. Introduction

Globally, coastal flooding have been devastating events causing cost for human environment, increase property damage, and around 20 million people are exposed to present high tide levels and 200 million to storm tide levels [1,2]. Currently, the Intergovernmental Panel for Climate Change (IPCC) report suggests that the global mean sea levels will increase 36–71 cm by 2100 based on Representative Concentration Pathway (RCP) 4.5 mid emissions scenario [3]. This situation may increase the vulnerability of coastal regions, especially of cities, due to demographic trends and economic expansion [4,5]. Meanwhile, in developing countries where various types of agricultural activities dominated the local economies, the impact is either ignored or simplified using rough estimates because of low expected losses [6–8]. Moreover, local types of agriculture such as solar salt production in tropical countries are also facing the impact of tidal flooding in particular location, which is overlooked in the global discussion. This type of agriculture, however, has the potential to generate revenue from salt in various aspects, not only in terms of salt product quality, but also for tourism, or even partly for coastal research centers [9].

Currently, solar salt production is acknowledged to be a marginal economic sector, especially in Indonesia [10]. It manually operates through traditional technology by using solar evaporation [11], locally referred to as *'maduranese'* method. The process starts in the saltpan. Seawater is let into the first and largest concentrating pond, or concentrator, through an inlet [12]. Most of the salt farmers are producing sea salt only during the dry season (April–October). The timing of the process highly depends on weather conditions, as rain reduces salinity and clouds decelerate evaporation [13]. Tidal flooding upon solar salt farming areas frequently occurs during high tide in the production period and thus threatens the production and distribution processes (see Figure 1). Between the 5th and 8th of June 2016, high tide flooding inundated hundreds of hectares of salt ponds in Cirebon, West Java, which is one of the major producer sites for salt in Indonesia [14,15]. Concurrently, the similar astronomic phases during the inundation events, showed the increase of high water level and low water level [16]. Temporarily, there was another nuisance flood between 23rd and 25th May 2018 along the north coast of Java (locally referred to '*Pantura*') during high tide [17]. Both events were narrated widely in both electronic and printed media.

Figure 1. Setting of traditional solar salt production area including: (**a**) salt evaporation pan and channel; (**b**) nearest pond to the sea; and (**c**) inundated pond due to high tide during fieldwork on 7 January 2018, 11:00 UTC (17:00 local time).

Tidal hydrodynamics in the Java Sea are complicated, due to their rough shallow bottom topography, diverse types of coastlines, and the interference of tidal waves propagating from the Pacific Ocean, Indian Ocean, and South China Sea [18]. Koropitan and Ikeda [19] previously investigated the implication of the barotropic tides into four tidal harmonic constituents using a three-dimensional (3D) hydrodynamic model combined with observation data and have suggested that the semi-diurnal M_2 component dominates over Java Sea. Several studies show that wind factor has a minimum contribution to tidal propagation in Java Sea [18–20]. Tidal flooding in the northern part of Java periodically rises in July and August during the East Monsoon period [21]. In recent years, the tidal inundation comes not only at high tide but even at the regular tide in some areas along *Pantura* [22]. Furthermore, the local economy, such as salt production which is dependent on coastal conditions, is eventually disrupted during these events.

Tides caused by the gravitational effects of sun and moon are periodic and very predictable [23–25]. Tidal floods (also defined as "nuisance" flooding) are occurring more often during seasonal high tides or minor wind events, and the frequency is likely to escalate intensely in the forthcoming decades [26]. Currently, the impact of tidal flood is usually modeled using planar approach in geographic information analysis. This approach assumes areas lower than a particular elevation to be inundated utilizing digital elevation model (DEM) and geographic information system (GIS) [27,28]. Geophysical processes including bottom friction or motion transfer are not considered in these particular models [29]. Ultimately, the uncertain behavior of the coastal system during a coastal flooding event is still a challenge in this model [30]. Previous work on GIS modeling presents various resolutions of DEM, which suggest using high resolution of elevation data to increase accuracy [24,31,32]. However, it is the hydrodynamics of the tide that is responsible for the size of the tide range, high and low waters momentum, and tidal characteristic, as well as the speed and timing of the tidal current [33]. An integrated approach considering both aspects is recommended, particularly for smaller areas and in cases where details are essential [29,34,35].

In this study, a simulation of the tidal flooding on salt production area is presented. Furthermore, investigations of high tide flooding in the salt production area of Cirebon are not available. The method, implemented in past events using hydrodynamic model with additional inputs on behavior of the coastal system during a tidal flooding event. This approach retrieves the values of the flood impact on the parcel of salt pond from two-dimensional (2D) (floodplain flow) tidal inundation simulation. Against the above background, the objectives of this study are to: (i) validate the tidal flooding that happened in June 2016 and May 2018 using a hydrodynamic model; (ii) analyze the highest tidal elevation and factors associated with flooding using tidal constituent and wind; (iii) plots tidally inundated area upon solar salt production area by considering the spatial distribution and the depths. Finally, this research offers better accuracy of analysis on the distribution of tidal flooding in salt production areas within limitation of tidal flood data.

2. Location of the Study Area

Cirebon is located 6°30′—7°00′ S and 108°40′—108°48′ E. It covers an area of 990.36 km². Administratively, Cirebon is a part of the West Java Province and is bordered by the Java Sea, and by Indramayu in the north, Kuningan in the south, Central Java Province in the east, and Majalengka in the west (Figure 2). It is a typical lowland with an average elevation of 0–25 m and covers 64,500 hectares. Cirebon connects the capital city of Jakarta with major cities in central and east Java. Cirebon has 40 districts, 424 villages, and 12 sub-districts. Based on BPS [36], the port city has an approximate population of 2.1 million, with 2205 inhabitants per km² and a population growth averaging of 1.28% per year.

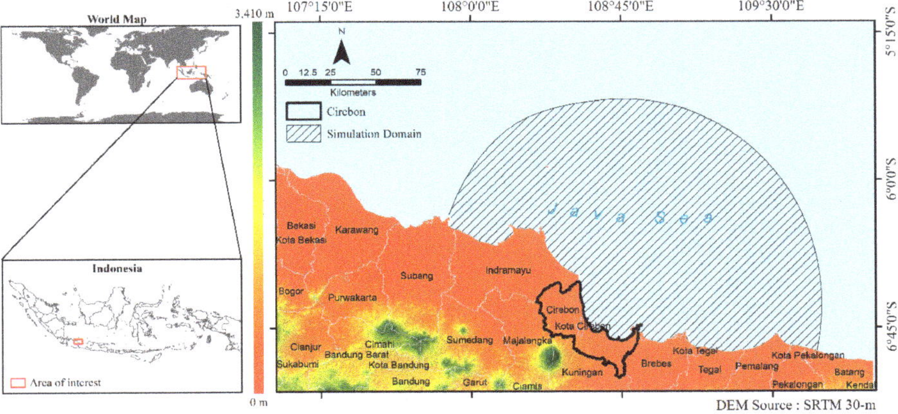

Figure 2. Geographical situation in Cirebon within typical coastal lowland adjacent to Java Sea and simulation coverage area.

Along with agriculture, salt production is shaping the local economy in the coastal region of Cirebon. The salt ponds cover 7819.32 hectares and provide jobs for 3707 people, such as pond owners, salt workers, and intermediaries [37,38]. Salt production predominantly takes place in low-lying areas dominated by alluvial deposits alongside with mangrove ecosystems. The salt production period in Cirebon begins during southeast monsoon. Most of the farmers start to store seawater in April, May, or June depending on the weather. They start to collect the brine daily and generate yields 0.5-1 ton/hectare/day. The dry season begins in March and ends in September, with a mean temperature of 32.8 °C, while the rainy season usually lasts from October to February, with an average rainfall of 1300-1500 mm/year and an average temperature of 24.2 °C [36]. The tidal regime is dominated by a mixed semidiurnal type and experiences two high and two low tides of different scales each lunar day. This tidal characteristic dominates the tidal cycle along Java sea [39].

The Java Sea is mainly identified as shallow water within roughly rectangular morphology, a mean depth of 50 m, a length of 950 km, and a width of 440 km [19,40]. The tidal range in the Java Sea is approximately 1.2–2 m, with peak values around Surabaya, Madura, and Bali [41,42]. The Java Sea is strongly governed by the monsoon climate. The northwest monsoon (NWM) reaches its peak between December and February (DJF) and it is usually characterized by frequent rainfalls and windy periods, while the Southeast monsoon (SEM) extends from June to August (JJA) and is usually characterized by much lower rainfalls [43].

3. Materials and Model Description

3.1. Data Acquisition

This study has used several data to simulate post-events of tidal floods. Firstly, the bathymetry and land topography information for the domain areas were handled as two main inputs for the model. Bathymetry of the Java Sea has been generated using gridded national bathymetry of Indonesia (BATNAS) provided from the Geospatial Information Agency (we referred to BIG: '*Badan Informasi Geospasial*' in Indonesian) (http://tides.big.go.id/DEMNAS/) within a 6 arc-second resolution. This data has been produced through the inversion of gravity anomaly of altimetry by adding sounding data carried with single and multi-beam surveys, which has better resolution in coastal areas than GEBCO (30 arc-second) [44,45]. Land topography data was resolved using the DEMNAS (0.27 arc-second resolution) also from BIG. DEMNAS is national seamless digital elevation data which already constructed within assimilated data of IFSAR (5-m resolution), TerraSAR-X (5-m resolution) and ALOS PALSAR (resolution 11.25 m), by adding stereo-plotting mass-point data [45]. This research draws on a previous approach by Tehrany [46] and Zalite [47] to utilize detailed topographic data for flood models. Here, five subsets (1309-12, 1309-14, 1309-21, 1309-23, and 1309-24) of DEM within the 0.27 arc-second spatial resolution were employed, and merged into a single raster data using GIS.

Secondly, tidal level data for Cirebon waters were captured from local tidal station in Cirebon port operated by BIG. In this case, we used hourly data of both selected period of simulations. As additional input, the simulation included wind data in the form of wind velocity, and wind direction of the Jatiwangi station. This data was extracted from OGIMET online meteorological database (https://www.ogimet.com/). Lastly, this model involved the latest (updated 2015) of the salt parcel dataset taken from the ministry of marine and fisheries affair during the salt inventory mapping project together with the national geospatial agency and PT. Garam [38]. At this point, salt parcel can clearly separate each pond by scale of 1: 15,000 in polygon format and was referenced into the World Geodetic System 1984. Details of data needed on this research are mentioned in Table 1.

Table 1. Data requirement for salt farming impact due to tidal flood.

Data Type	Resolution	Location	Period	Source
Bathymetry	6-arc"	Modeled expanse	2018	BIG
Topography	0.27 arc"	Cirebon area	2018	BIG
Water level (ζ)	1 h	Cirebon port	June 2016, May 2018	BIG
Wind velocity (u, v) and direction	1 h	Jatiwangi	June 2016, May 2018	OGIMET
Tidal calibration (ζ, u, and v)	1 h	T. Sari, Pangenan, Bungko	June 2016, May 2018	TPXO, TMD
Salt parcel	1:15,000	Cirebon area	2015	BIG

3.2. Model Setup

This research applied a numerical hydrodynamic model (HDM) to forecast run-up and tidal inundation in the salt production area in Cirebon. HDMs originated through resolving Laplace Tidal Equations and using bathymetry data as boundary conditions [48]. The module of MIKE 21 package was used, as one of the most widely used hydrodynamic model in computation by Danish Hydraulic Institute (DHI), including the assessment of hydrographical sequences in non-stratified waters, coastal flooding and storm surge, inland flooding, and overflow [49–51]. The MIKE package also represents user-friendly GIS interfaces and provides better possibilities to simulate the flooding using elevation data and bathymetry [52].

The model employed MIKE 21 Flexible Mesh (FM) to simulate water levels and tidal floodings in selected events. These tools were utilized using the input of tidal gauge records to indicate the spatial variability of tidal flood characteristics of two events. The model was applied for two separated months, June 2016 (event A) and May 2018 (event B), which covered the occurrence of the selected tide flood events. The unstructured triangular mesh with 87,103 nodes and 170,501 elements was generated in the simulation and covered 11,515.20 km^2 (see Figure 3). The mesh file in ASCII format included information of the coordinates and bathymetry for each node point in the mesh [50]. The grid dimension differs by 2800 m in the northeast ocean boundary, the smaller grid size around 450 m in the outland, and 120 m in the inland area.

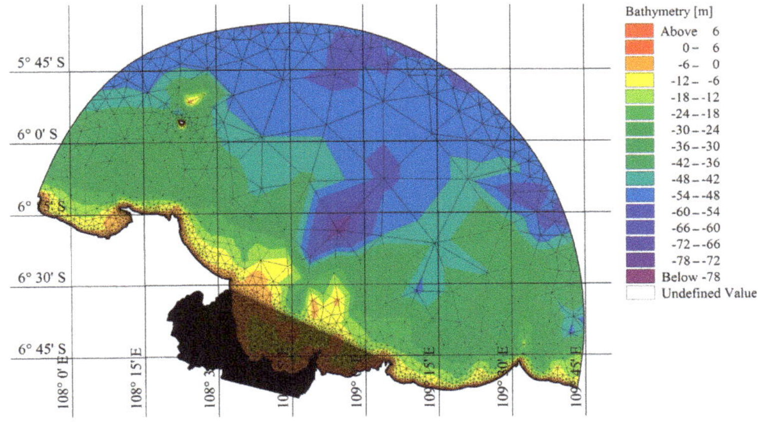

Figure 3. Mesh generation of Cirebon waters and part of Java Sea, bathymetry (in meters), comprising of 87,103 nodes and 170,501 triangular elements.

The tidal flood events, which were triggered by high tides, were simulated through forcing tidal elevations at open borders, winds, and temperatures. The tidal height was calculated by hourly local station measurement using a harmonic approach. This classical harmonic analysis represents the tidal forcing as a set of spectral lines, demonstrating the predetermined set of sinusoids at specified frequencies [53,54]. This stage resulted in nine tidal components (M_2, S_2, N_2, K_2, K_1, O_1, P_1, M_4, and MS_4) that correspond to specific physical phenomena such as the period of the moon around the earth or friction against the seabed in shallow seas.

Following step of this model was to include the wind energy and set tilt facility to declare water level correction along the boundary of the waters using Navier-Stokes equations. These equations deliver an appropriate model for wave overtopping and overcome sophisticated hydrodynamics, including wave breaking and its theoretical limitation [55,56]. Furthermore, the flood and drying (FAD) ability of this model assisted the water run-up simulation and executed the inundation process of high tides. This scheme has been alternated to describe the coastal situation, where it can be flooded at one time but dry at other times. This study use recommended value in Thambas [50], h_{dry} = 0.005 m, flooding depth h_{flood} = 0.05 m and wetting depth h_{wet} = 0.1 m.

It should be noted that the elevation data is used in this simulation without considering water surface evaporation. Due to the high complexity of the site study and the limitations of the model, the following steps were considered for the tidal flood simulation: bottom friction is based on Manning's approach, with the ranges of friction coefficients from 40 for water to 32 for land [57]. A Manning number in the range 20-40 m $^{1/3}$/s is typically applied with an advised value of 32 m $^{1/3}$/s if other data is unavailable [58]. The Manning number relates to the flow path and peak time of flooding and does not have a significant effect on flood distribution and depth [59]. Furthermore, the simulation included horizontal eddy viscosity using the Smagorinsky type within a value of 0.28 [49].

After the work with MIKE had been completed, the result was exported using "MIKE2Grid" for further spatial processes. This produced an ASCII file that is readable in ArcGIS. Importing this file and reorganizing the classes produced the inundation map of MIKE 21. Finally, we superimposed the grid data to the salt parcel dataset and used the tidal simulation as a basis in inundation analysis process. This plot dataset has a shapefile format, which is suitable for further handling in ArcGIS. The inundation map was created by using grid data of both simulations in ArcGIS. The two-top water levels of the selected simulations, which were identified with the maximum value in the data series that was used as a benchmark for inundation analysis. In this research, certain assumptions were made, such as that no precipitation data inputs were used during the period of the incident as it may raise the inundation level, no sea-level rise and land subsidence were considered in the simulation, as there is still no strong local evidence of both factors in the research location. The overall steps of this research are summarized in Figure 4.

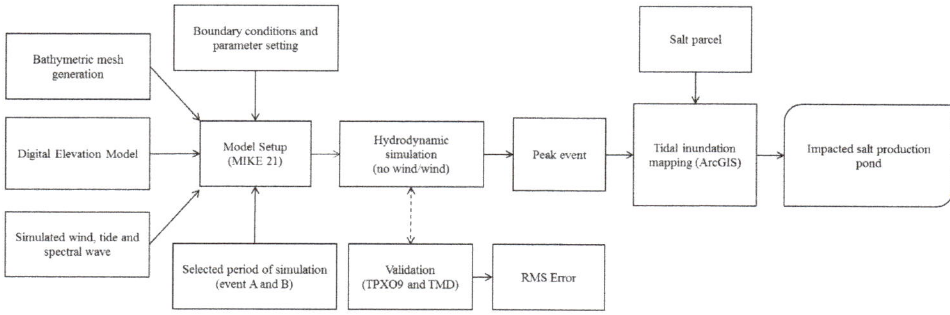

Figure 4. Diagram of simulation of tidal flood and impact mapping procedure including validation process.

4. Results

4.1. Validation of Tidal Simulation

To get a sound validation of our model for tidal simulation, this study used tidal data of the open boundary model that were obtained from the global tidal model. Following the step from Ningsih et al. [60], the model compared the results of simulated sea level from MIKE 21 with tidal station in Cirebon from BIG and also the global tidal model TPXO9 (this model can be accessed on http://volkov.oce.orst.edu/tides/global.html). TPXO9 is the latest version of TPXO-series [61,62], which includes global tidal solutions with 1/6° resolution that fit, in least-square, both the Laplace's equation and also the long track averaged data from TOPEX/Poseidon and Jason (on T/P tracks since 2002) [61,63]. Tidal constituents of the tide record from BIG tidal station perform comparable values with tidal constituents from the global tide modeling TPXO in Indonesian waters [64]. At this point, the wind factor was excluded and focused on gravitational force only. Moreover, the tidal current velocity was verified with Tidal Model Driver (TMD). This free MATLAB package offers harmonic constituents for tide models, making predictions of tide height and also currents [65]. Verification points of the selected simulations (event A and B) are located in Tawangsari, Pangenan, and Bungko (as pointed P1-P3 in Figure 5). The model exposed the statistical correlation using the Pearson value (r) of the three locations with general tidal model of TPXO9, and presented the value of the Root Mean Square (RMS) error. The RMS error was calculated with:

$$x_{RMS} = \sqrt{\left(\sum_{i=1}^{n} x_i^2\right)/n}$$

(1)

where x_i is the ith point of the chosen area, were calculated in the region 6°S–7°S, 108°E–109°E (the Java Sea). The overall locations verified excellent correlations between simulated outcomes and those of TPXO9 and TMD.

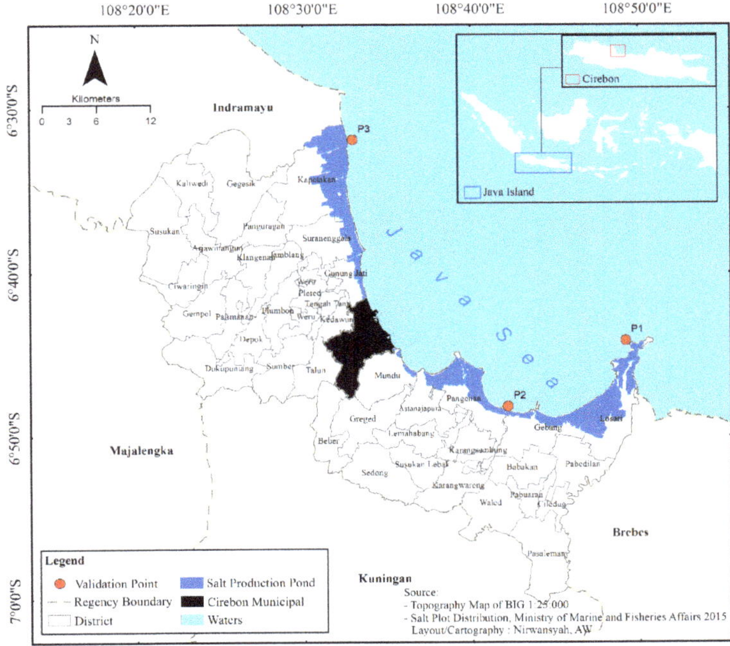

Figure 5. Salt production area for tidal simulation and validation points (P1-P3).

The Pearson correlation of the tidal height simulation is in the range of 0.903–0.908 for event A and of 0.848–0.903 for event B with the RMS Error within approximately 0.069–0.100 m. For the u-velocity component, the correlation shows a coefficient around 0.833–0.965 with RMS Error of about 0.023–0.0196 m/s for event A. For event B, the correlation coefficients range between 0.570 and 0.877 with a RMS Error around 0.019–0.190 m/s. Furthermore, the v-velocity component shows a good agreement of the correlation coefficient, whereas the number of the correlation is about 0.683–0.824 with RMS Error about 0.040–0.061 m/s. Although there were some inconsistencies between u-velocity components in MIKE 21 simulation on event B, overall, the simulation results managed well with the TMD data. Lower values of RMS Error suggested the appropriate model to the data points; likewise, values of Pearson close to the maximum point of one (value of 1) indicated that the model has a strong correlation to the water level data [66,67]. Detailed values of Pearson Correlation and RMS Error between simulation and global tide model of TPXO9 for water elevation and those TMD for tidal velocity elements are shown in Table 2. The illustration for tidal height (ζ), u and v-velocity at verification points can be seen in Figures 6–8.

Table 2. Pearson Coefficient (r) and Root Mean Square (RMS) Error between simulation rates in MIKE and TPXO9 for water elevation and velocity components from TMD.

Stations	Tidal Height (ζ)		u-Velocity Component		v-Velocity Component	
	r	RMSE (m)	r	RMSE (m/s)	r	RMSE (m/s)
Tawangsari						
• Event A	0.903	0.071	0.894	0.0196	0.715	0.061
• Event B	0.891	0.075	0.877	0.0190	0.683	0.059
Pangenan						
• Event A	0.904	0.100	0.833	0.029	0.724	0.050
• Event B	0.903	0.075	0.814	0.019	0.755	0.059
Bungko						
• Event A	0.908	0.069	0.965	0.023	0.824	0.048
• Event B	0.848	0.088	0.570	0.073	0.705	0.040

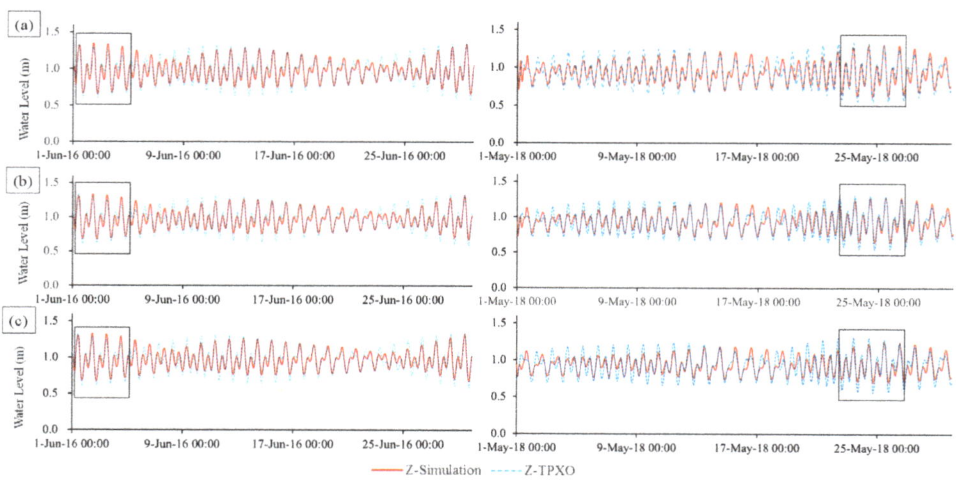

Figure 6. Comparison of water level between simulation and TPXO on June 2016 (left) and May 2018 (right) at (**a**) P1—Tawangsari; (**b**) P2—Pangenan; and (**c**) P3—Bungko.

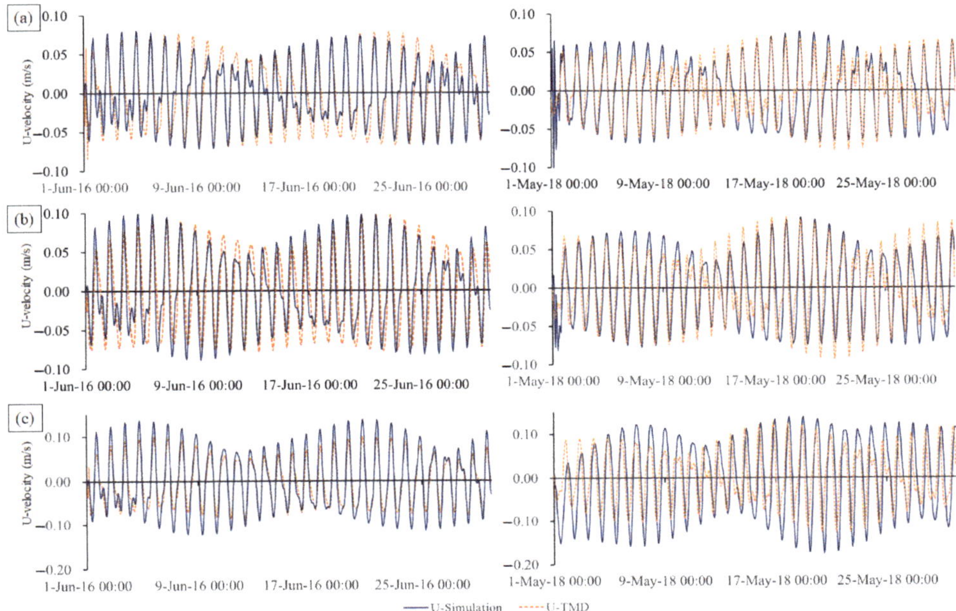

Figure 7. Comparison of *u*-velocity component between MIKE and TMD on June 2016 (left) and May 2018 (right) at (**a**) P1—Tawangsari; (**b**) P2—Pangenan; and (**c**) P3—Bungko.

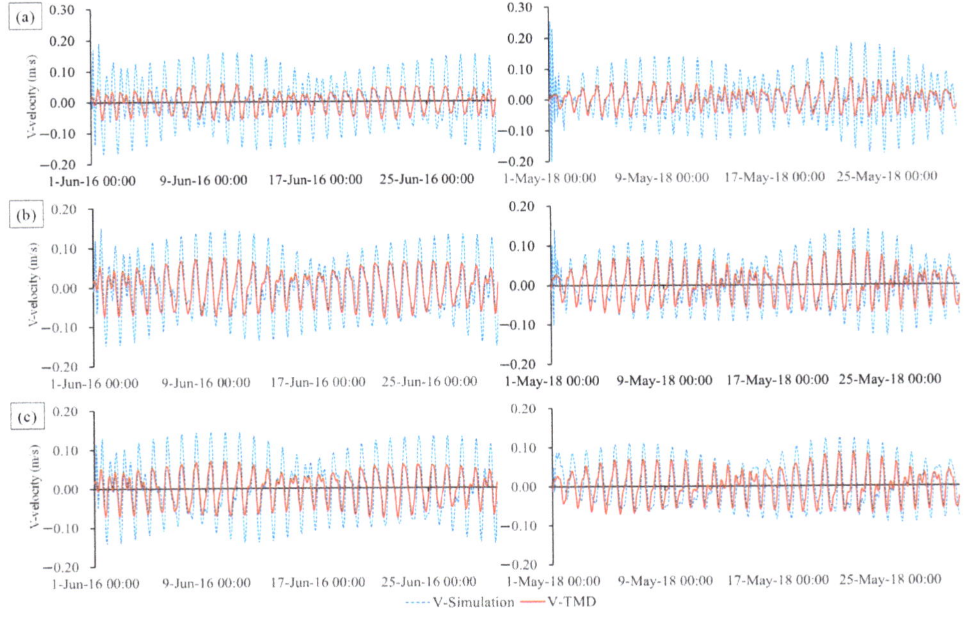

Figure 8. Comparison of υ-velocity at verification points and TMD on June 2016 (left) and May 2018 (right) at (**a**) P1—Tawangsari; (**b**) P2—Pangenan; and (**c**) P3—Bungko.

Spring tides (at the new moon phase) appeared in between the flooding events of event A, which was recorded from June 1st–5th, 2016. In event B, the high tides were tracked from the 23rd to the 28th of May during the end of the moon phase. In both of these conditions, the rise of waves is a natural phenomenon due to moon force (M_2) [20]. However, on the three tide validation points, the increase of high water levels (HWL) and low water levels (LWL) during the inundation events can clearly be compared to the same astronomic phases tides data before and after the events (see black boxes in Figure 6 above). Moreover, the horizontal (u) velocity on three sample locations show typical performances within range of 0-10 m/s and vertical (v) velocity in range of 0-28 m/s (see Figures 7 and 8). Here, it can be seen that there are differences velocity level between the simulation and the TPXO model in three sample locations, which are explained in the previous section.

In the next step, the wind air pressure data were engaged in the model as an additional factor in tidal propagation. Wind data from OGIMET has been entered into the simulation for both selected periods. Adding hourly wind data into the model shows minor differences of amplitudes and phases of tidal constituents. The simulations show that there is an insignificant difference in the tidal pattern due to the relatively small effect of wind, as the velocity of wind dominantly emerged from the north with a maximum speed of 6.8 m/s during event A and an average velocity around 0.81 m/s. For event B, an extreme increasing velocity of 60.48 m/s is recorded in the simulation. Ultimately, the average wind speed is approximately 0–1.49 m/s. Here, the assumption has been made that typically calm wind in both selected periods has a minor impact on tidal floods along the coast.

Wind velocity confirmed a minimum correlation to the water level (Pearson correlation 0.1902-0.1905 for event A and −0.021 to −0.031 for event B). The negative correlation probably relates to the minimum velocity of the wind, as OGIMET provides hourly datasets in both periods of simulation. The typically calm wind during these periods provides a better situation for evaporation in the salt production process. Nevertheless, high tide continually increases the potential of inundation in coastal areas where salt production takes place. At the same time, the water elevation of the simulation also shows significant correlation to the observation data from tide gauges (Figure 9a,b). Figure 9c,d present the peak tide levels for both events. Here, the series of surface water (z) data from MIKE 21 simulation are also used to determine the nine tidal components. Here, Table 3 presents the variability of tidal constituents in both periods of the model.

Table 3. Tidal amplitudes constituents in Cirebon resulted from simulation.

Simulation	Tidal Constituent									
	Z_0	M_2	S_2	N_2	K_2	K_1	O_1	P_1	M_4	MS_4
Event A	0.939	0.143	0.062	0.044	0.051	0.107	0.040	0.042	0.001	0.002
Event B	0.902	0.143	0.059	0.048	0.004	0.088	0.049	0.014	0.004	0.001

Based on the calculation, the wind involved in the simulation showed correlation values to tidal gauge observation 0.761 with RMSE of 0.134 m for event A and 0.79 with RMSE 0.120 m for event B. As a result, surface elevation at the peaks of both tidal events in Cirebon reached 0.38 m (event A) on the 2nd of June 2016 12:00 UTC and 0.40 m (event B) on the 25th of May 2018 11:00 UTC. Gurumoorthi and Venkatachalapathy [68] and Pugh [69] mentioned that the relative importance of diurnal and semi-diurnal components differ with geographical position and can be calculated by the formulation factor:

$$F = (O_1 + K_1)/(M_2 + S_2) \tag{2}$$

In Equation (2), the average constituents confirm the typical mixed, predominantly semi-diurnal tide within 0.73 and 0.68 for both events A and B. Considerably, the amplitudes of semidiurnal tidal constituents were higher than the diurnal tides.

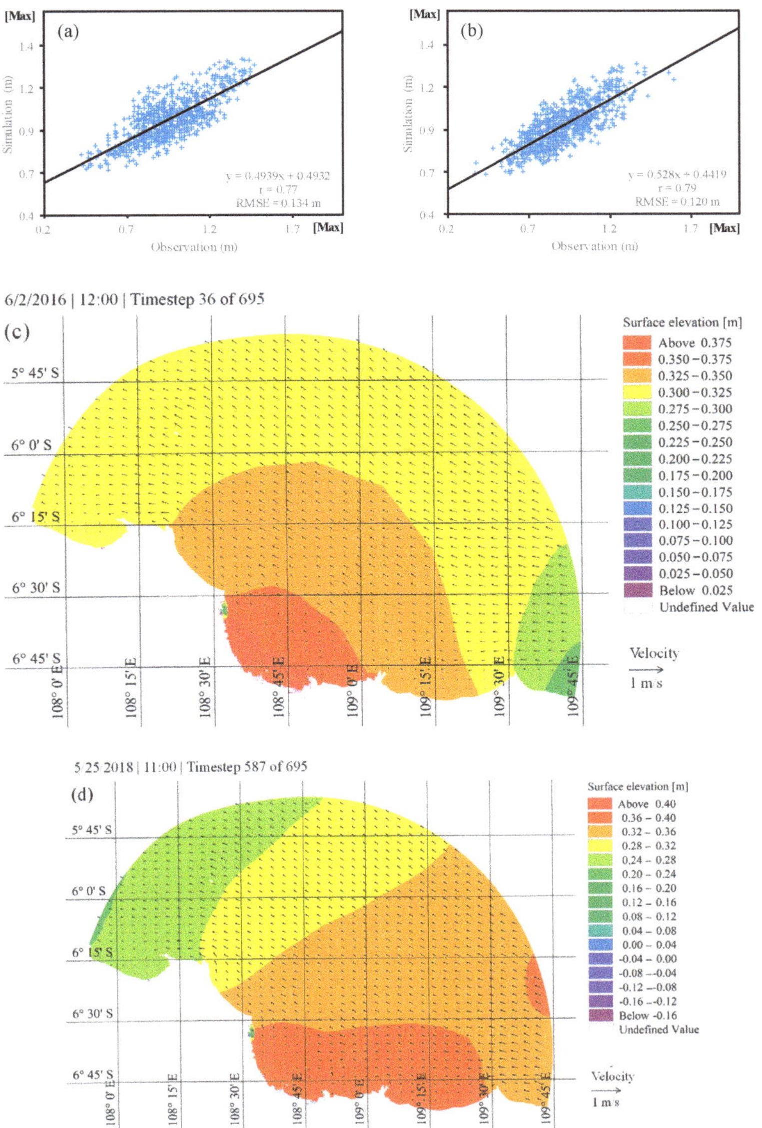

Figure 9. Scatterplot and RMS error of simulated surface water elevation with tide gauge observation on: (**a**) event A; (**b**) event B within peak level of water during simulation on; (**c**) 2 June 2016 12:00 UTC (steps 36); and (**d**) 25 May 2018 11:00 UTC (steps 587).

4.2. Maximum Tidal Height and Exposed Salt Production Area

Understanding the impact of tidal flood dispersal on the coastal area demands a model of inundated area caused by tide water level [24]. Equilibrium flood mapping or the "bathtub" approach compare the maximum total water level and ground height. At those places, where the land is lower than the expected maximum water level, it will be flooded [70]. The expected water depth for each salt pond can have major implications for tidal management, especially for vulnerability measurements as damage is often associated with the depth of inundation and its duration. The simulation showed that both tidal floods were forecasted to be generated by meteorological factors. Here, M_2 tidal response provides the dominant influence (which both events record 0.143) in amplitudes of Cirebon waters. As a result, the surface elevation during the maximums of both tidal events have been exported in ArcGIS using Mike2Grid tools and visualized water level and the spatial distribution of the inundation upon salt production area (see Figure 10).

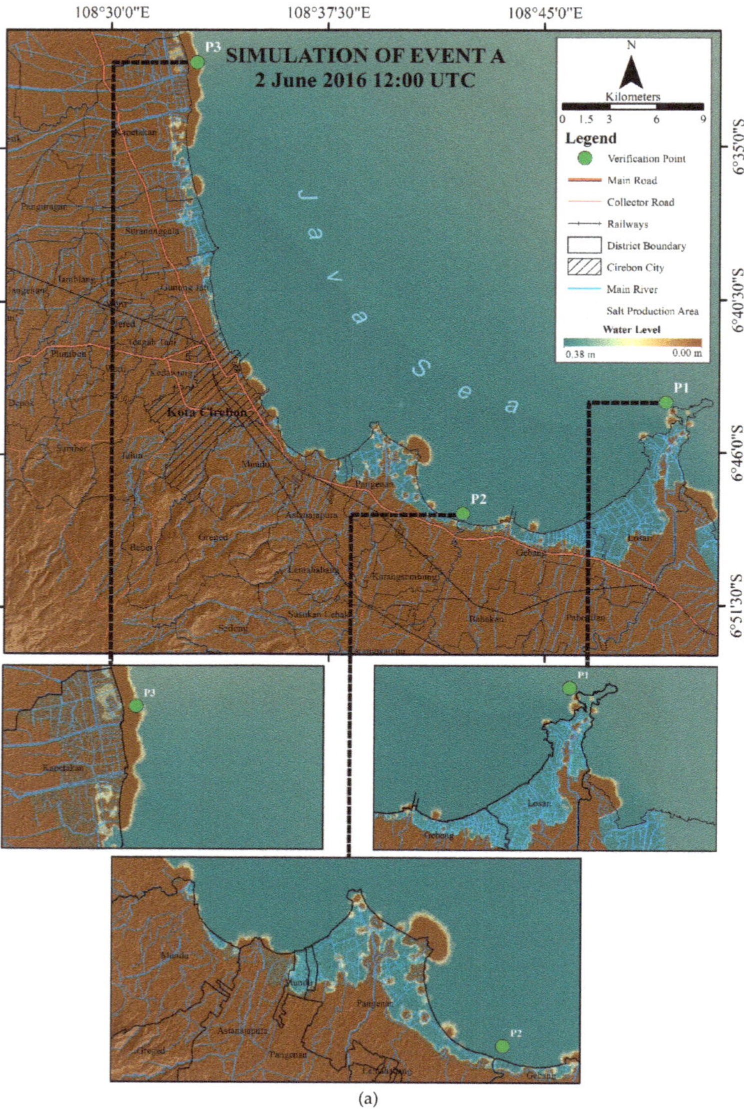

Figure 10. *Cont.*

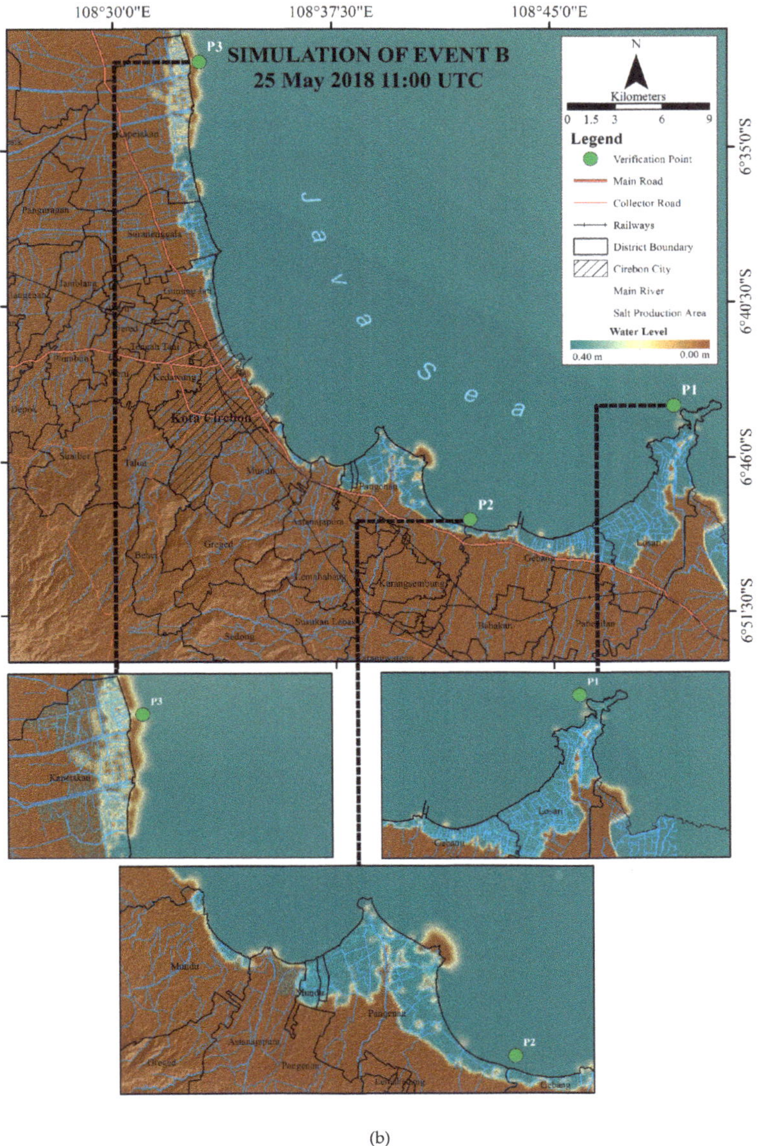

(b)

Figure 10. Simulation of water level during peak period of: (**a**) event A and (**b**) event B.

The simulation map shows that tidal inundation occurs along the coastline of Cirebon during peak water levels. The grid data was superimposed with detailed DEMNAS to investigate the impact of tidal flooding upon solar salt production land. A reclassification process in GIS elaborates the tidal dynamics and flood depth upon salt pond in the study area. Here, each salt parcel has a single value of depth level through spatial joint between both vector types of water level and parcel of salt pond datasets (Figure 11). Thus, this results in the appropriate value for inundation for each pond that has been impacted by the tidal occurrence for both events.

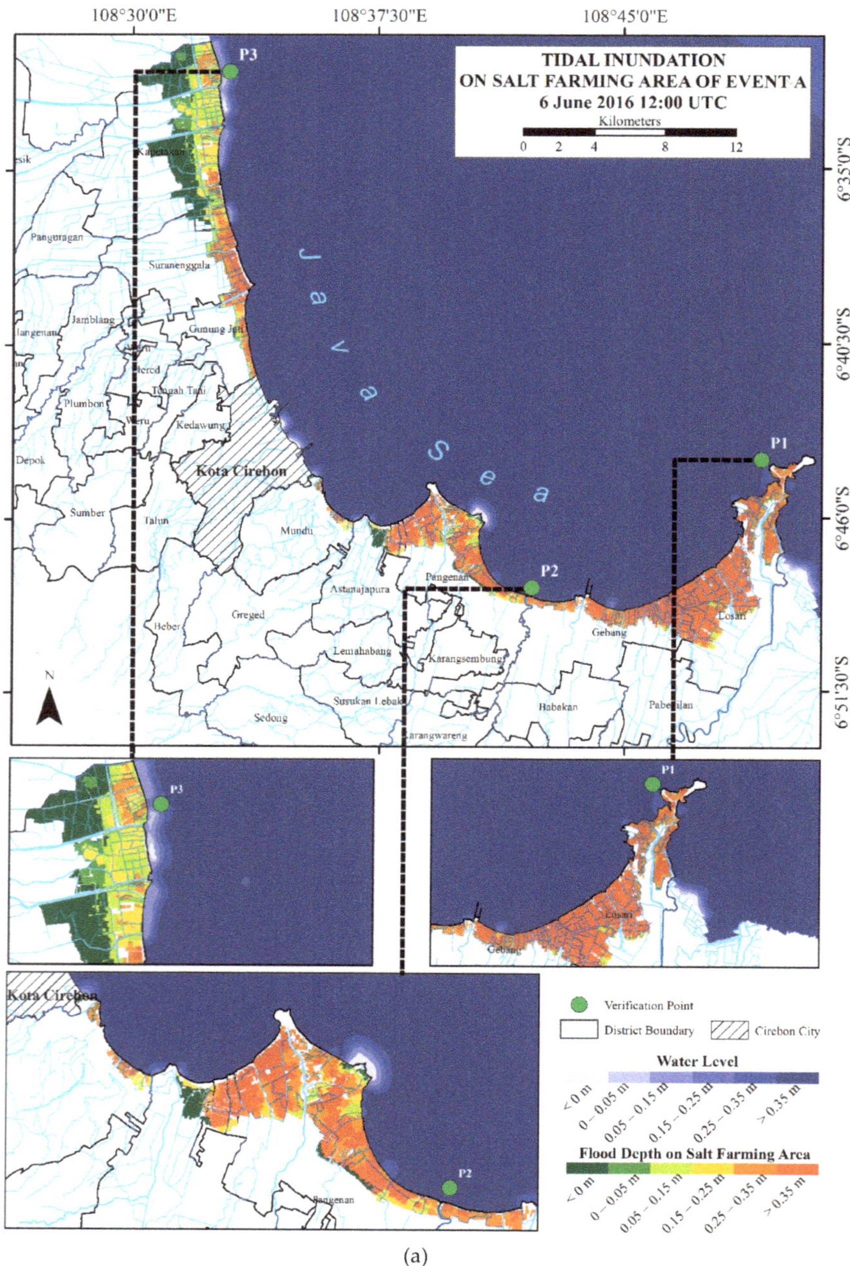

Figure 11. *Cont.*

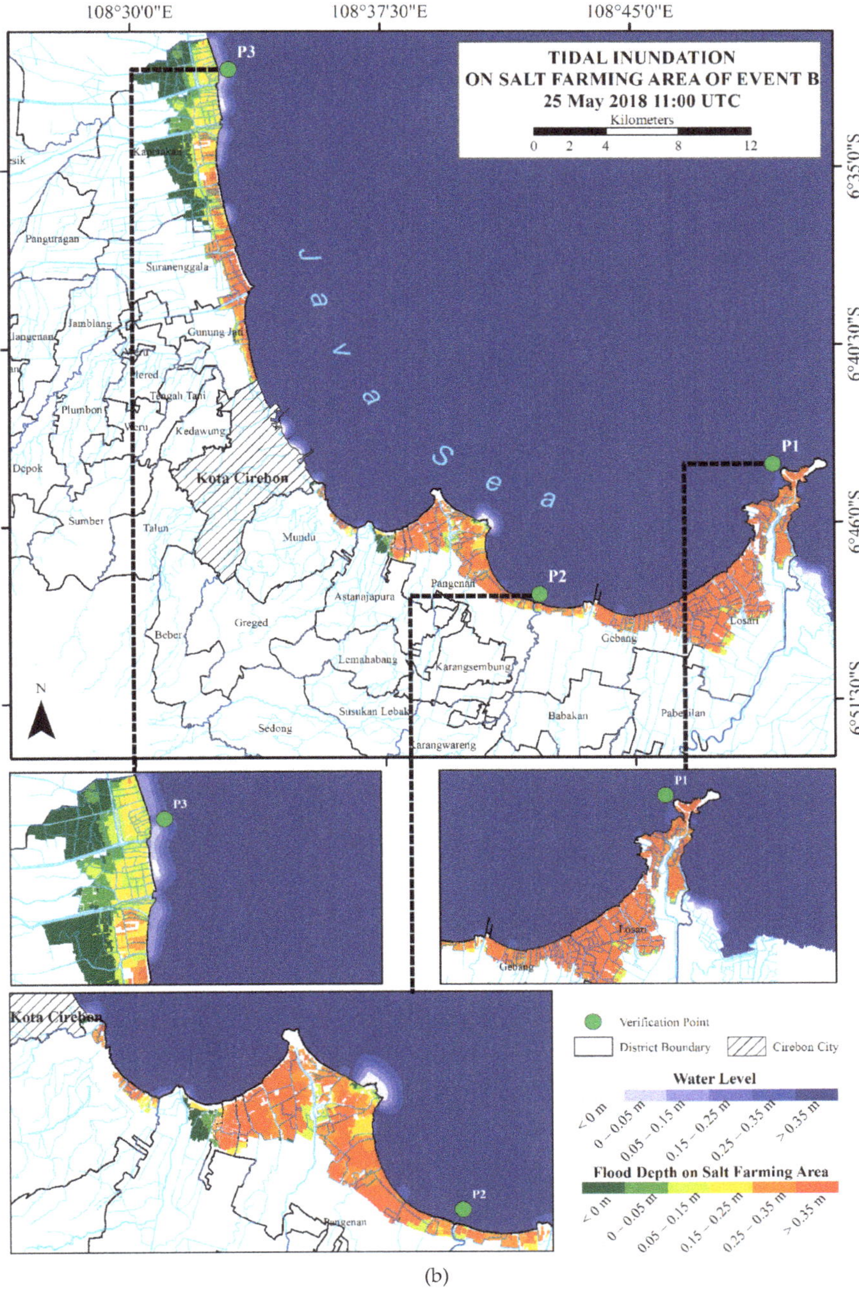

(b)

Figure 11. Estimated inundation level for each pond of during the highest tide of (**a**) event A and (**b**) event B (each parcel contains single value of inundation depth).

As mentioned in the previous section, the results regarding the submerged area of around 0–0.38 m for (event A) and 0–0.40 m for (event B), have significantly affected the salt production areas in Cirebon. Based on previous maps (Figure 11), it can be seen that around 1990.55 ha of salt production pond in Losari were inundated during event A and 1992.07 ha during event B. This district is also recorded as the most impacted area due to both tidal flood events (99.92% and 99.99% of total cultivated area in Losari). The salt production area of Gebang, which is located in the west part of the study area, has also been flooded up to 816.32 ha (100%) during the events A and B. At the same time, a slight increase of flood coverage has occurred in Kapetakan due to both tidal events. During the peak level of event A, almost 56.15% or 1,538.96 ha were exposed to tidal floodings, and 57.22% or 1,568.34 ha suffered inundation according to our simulations. In the middle part, tidal heights of both selected simulations in events A and B have submerged Suranenggala, Gunungjati, Mundu, and Astanajapura to a lesser degree in terms of total area, but with more significant percentages of inundated salt pond (approximately 49–99% in both events). The model presents the areas of inundation on A and B events as it is drawn in Figure 12.

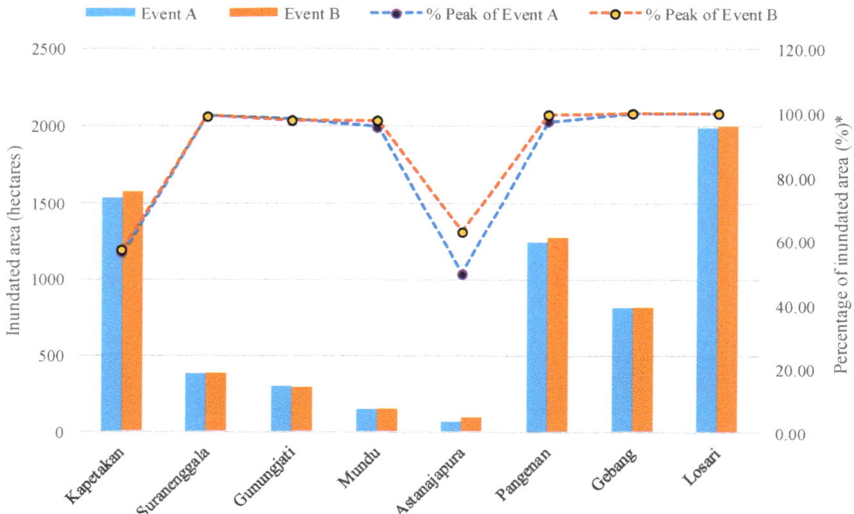

Figure 12. Tidal flood accumulative distribution area of inundation in Cirebon due to high tide at event A and event B.

This model shows relatively small differences in terms of the affected area during both simulation periods, as the wind factor has less impact and relatively similar water level. The simulated inundated salt production areas for A and B peak events are estimated to be 6489 ha and 6570 ha, respectively, which equals to 83 % and 84.2 % of the total salt production area in Cirebon. Overall, the peak depth of > 0.35 m dominates the tidal flood sequence, with 41.9% and 45.5% of the area being inundated to such a degree. This depth level has a significantly larger effect in destructing dikes compared to a lower flood level. Meanwhile, around 16–17 % of ponds are relatively safe from these events, as they are located further inland on higher elevations. The less impacted area is estimated within >0-5 cm depth, which covers 1.7% (event A) and 4.1% (event B) of the total inundated area. During the flood, salt production was postponed and stopped until the water receded. It has to be noticed that higher flood levels also take a longer time to recede, which prolongs the preparation and the pre-production process, thus worsening the impact of the floods upon the salt production. Estimations of the salt production area that have been inundated based on simulation is presented in Table 4.

Table 4. Estimated areas and percentage of salt production in Cirebon affected by tidal flood in selected events (area in hectare).

Water depth	Inundated	Event A	Event B
> 0–5 cm	Area	132.4	317.2
	%	1.7	4.1
5–15 cm	Area	894.3	847.4
	%	11.4	10.8
15–25 cm	Area	722.6	687.9
	%	9.2	8.8
25–35 cm	Area	1463.1	1156.1
	%	18.7	14.8
> 35 cm	Area	3277.1	3561.4
	%	41.9	45.5
Total Inundated		**6489.4**	**6570.0**

5. Discussion

This paper presents simulation of the inundated areas upon salt farming due to tidal flood events. Tidal floods that occurred upon salt production area were triggered by the high tide events in Java Sea. That two events were similarly situated to tidal incidents located along the southern part of Java adjacent with Indian Ocean [71,72]. Both periods studied present similar tidal elevations. Inundation dominantly occurred in the western and eastern parts of the region. The total impacted salt production area in both events were about 6489.4 ha (event A) (83%) and 6570 ha (84%) (event B). As illustrated by Châu [73], the inundated area may be overrated based on data characteristics and the methods employed. Different resolutions of DEM may result in different total area of inundation. Furthermore, bathymetry, wind velocity, and Manning coefficient also correlate with the hydrodynamic process of tidal forcing.

This method, which relies on tidal characteristics and hydrodynamic parameters, leads to a usefulness tidal flood mapping for salt production areas. This idea improves the marine environment evaluation through cost-effective technique and limited data collection in particular coastal regions [74], and improved flood forecasting [56]. Based on performance, the hydrodynamic simulation's high degree of confidence with the global tide model can be placed as input to identify inundated area during tide events. The results express the significance of gaining reliable datasets into calibration and validation processes [75]. The availability of spatial data for the study area, including DEM, bathymetry, meteorological data, and salt parcel area, also give beneficial support for the models. Although wind data do not confirm significantly in the simulation performance, there were secure connections between our models with tidal records from local station (see Figure 9).

In this study, DEMNAS (0.27 arc-second) was the highest resolution elevation data available in Cirebon and representative for the simulation and tidal inundation mapping; however, more accurate results would be achievable through higher resolutions [34,76,77] such as LiDAR-derived DTMs [56,78], and more extended tidal gauges data. Previous work by Seenath et al. [34] acquired 10-m DEM in the flood modeling component and delivered relatively higher RMS error. Additionally, smaller frequencies of simulation, i.e., 5-sec [79,80] and 6-min [81] will improve the model's stability. While there are advantages to use a hydrodynamic model for tidal flood mapping, the required computation time and the resolution of input data may limit its application in practice, especially for larger areas (compare Seenath et al. [34]). In this case, DEMNAS performs better resolution on water depth visually, but within a more extended handling time. This data was also previously exported into polyline format along with bathymetry in the preparation step. Processing simulations take a much longer time than calculating a general bathtub model. The hourly data during the 30-day period of simulation took almost 48 hours (using a standard PC with Intel I5 and 8GB of RAM).

As Cirebon salt production operates in a very traditional manner, the local salt farmers rely on daily harvests and the tidal cycle. Meanwhile, the ability to recover from disasters is far from sufficient. A comprehensive risk analysis of the salt production area is urgently needed to be better prepared to deal with the more prominent impacts of tidal floods on coastal areas. Tidal flood simulations have the potential ability to lead for better evaluations, including the potential damage loss on case-based analysis. This data will enable farmers and stakeholders to better respond to future hazards and to build capacity to improve the quality of livelihoods in the tidally flooded areas.

6. Conclusions and Future Works

This study has developed a method to identify the tidal flood impact in different types of agriculture areas in the coastline where the tide is generally forced by local factors, using hydrodynamic models. The model simulates typical aspects of tidal flood in Cirebon coastal regions, where lunar force dominates the domestic tidal properties during salt production periods. The method allows for critical identification points of flooding in the simulation, rather than using sets of scenarios. This study also underlines the main interest in the possible analysis of marginal agriculture along the coast, where the tidal hazard may continue in the future and would make a more significant impact. Meanwhile, there is still a limited number of studies that emphasizes the exposure of tidal hazard in local traditions economic activities. Tidal flood impact mapping can be beneficial to increase awareness of salt farmers to the flood occurrences. Additionally, the uncertainty and volatility level of this type of flooding is driving the local government to put more attention, especially for countermeasure planning and efficient mitigation strategies. Using higher-resolution DEM, such as LiDAR, and echo-sounder survey data for detailed bathymetry can lead to an improvement for tidal flood assessments accuracy. Although this particular model is initiated for local-case study, it is believed that this technique can be developed at a regional scale with data limitation. Finally, this research will be more substantial to include the benefit-cost (B/C) analysis of the post tidal flood events.

Author Contributions: Conceptualization, Anang Widhi Nirwansyah and Boris Braun; methodology, software, validation, formal analysis and writing-original draft preparation, Anang Widhi Nirwansyah; writing-review & editing, Anang Widhi Nirwansyah and Boris Braun; supervision, Boris Braun.

Funding: The authors would like to acknowledge financial support provided by Indonesia Endowment Fund for Education (LPDP) within code of LPDP: PRJ-115 /LPDP.3/2017 provided to Anang Widhi Nirwansyah, Institute of Geography (University of Cologne) for publication funding and Universitas Muhammadiyah Purwokerto, Indonesia.

Acknowledgments: This project has been part of doctoral research at the Institute of Geography, University of Cologne, Germany. The authors express their gratefulness to the anonymous reviewers for their valuable advice. Part of the data used for this paper, such as bathymetry, DEM, salt parcel on scale 1: 15,000, administrative boundary was made possible through the support of Indonesian National Geospatial Agency (BIG), and Indonesia Ministry of Marine and Fisheries Affairs (KKP). Finally, we would like to thank Junika A. Fathonah (UNDIP) for helping us during MIKE simulation, and salt farmers in Cirebon for valuable information during fieldwork.

Conflicts of Interest: The authors declare no conflicts of interest and the founding sponsors had no role in the design of the study; in the collection, analyses, or interpretation of data; in the writing of the manuscript; or in the decision to publish the results.

References

1. Church, J.A.; Woodworth, P.L.; Aarup, T.; Wilson, W.S. *Understanding Sea-Level Rise and Variability*; Church, J.A., Woodworth, P.L., Aarup, T., Wilson, W.S., Eds.; Wiley-Blackwell: Oxford, UK, 2010; ISBN 978-1-44-432327-6.
2. Marfai, M.A.; King, L.; Sartohadi, J.; Sudrajat, S.; Budiani, S.R.; Yulianto, F. The impact of tidal flooding on a coastal community in Semarang, Indonesia. *Environmentalist* **2008**, *28*, 237–248. [CrossRef]
3. Church, J.A.; Clark, P.U.; Cazenave, A.; Gregory, J.M.; Jevrejeva, S.; Levermann, A.; Merrifield, M.A.; Milne, G.A.; Nerem, R.; Nunn, P.D.; et al. *Sea Level Change*; Cambridge University Press: Cambridge, UK; New York, NY, USA, 2013.
4. Bouwer, L.M.; Bubeck, P.; Wagtendonk, A.J.; Aerts, J.C.J.H. Inundation scenarios for flood damage evaluation in polder areas. *Nat. Hazards Earth Syst. Sci.* **2009**, *9*, 1995–2007. [CrossRef]

5. Ward, P.J.; Marfai, M.A.; Yulianto, F.; Hizbaron, D.R.; Aerts, J.C.J.H. Coastal inundation and damage exposure estimation: A case study for Jakarta. *Nat. Hazards* **2011**, *56*, 899–916. [CrossRef]

6. Forster, S.; Kuhlmann, B.; Lindenschmidt, K.E.; Bronstert, A. Assessing flood risk for a rural detention area. *Nat. Hazards Earth Syst. Sci.* **2008**, *8*, 311–322. [CrossRef]

7. Brémond, P.; Grelot, F. Review Article: Economic evaluation of flood damage to agriculture—Review and analysis of existing methods. *Nat. Hazards Earth Syst. Sci.* **2013**, *13*, 2493–2512. [CrossRef]

8. Merz, B.; Kreibich, H.; Schwarze, R.; Thieken, A. Review article "Assessment of economic flood damage". *Nat. Hazards Earth Syst. Sci.* **2010**, *10*, 1697–1724. [CrossRef]

9. Rodrigues, C.M.; Bio, A.; Amat, F.; Vieira, N. Artisanal salt production in Aveiro/Portugal—An ecofriendly process. *Saline Syst.* **2011**, *7*, 3. [CrossRef]

10. Rochwulaningsih, Y. *Marginalization of Salt Farmers and Global Economic Expansion: The Case in Rembang Regency of Central Java*; IPB: Bogor, Indonesia, 2008. (In Bahasa)

11. Munadi, E.; Ardiyanti, S.T.; Ingot, S.R.; Lestari, T.K.; Subekti, N.A.; Salam, A.R.; Alhayat, A.P.; Salim, Z. *Info Komoditi Garam*; Salim, Z., Munadi, E., Eds.; Badan Pengkajian dan Pengembangan Perdagangan: Jakarta, Indonesia, 2016; ISBN 978-9-79-461890-5.

12. Antczak, K.A. *Entangled by Salt: Historical Archaeology of Seafarers and Things in the Venezuelan Caribbean, 1624–1880*; College of William and Mary: Williamsburg, VA, USA, 2017.

13. Helmi, A.; Sasaoka, M. Dealing with socioeconomic and climate-related uncertainty in small-scale salt producers in rural Sampang, Indonesia. *J. Rural Stud.* **2018**, *59*, 88–97. [CrossRef]

14. Metrotv 700 Hectares of Salt Pond in Cirebon Submerged Coastal Flood. Available online: http://m.metrotvnews.com/jabar/peristiwa/JKR4MYQb-700-hektare-tambak-garam-di-cirebon-terendam-banjir-rob (accessed on 29 September 2017). (In Bahasa).

15. Lia, E. Puluhan Ribu Ton Garam di Cirebon Tersapu Banjir Rob page-2: Okezone News. Available online: https://news.okezone.com/read/2016/06/17/525/1418255/puluhan-ribu-ton-garam-di-cirebon-tersapu-banjir-rob?page=2 (accessed on 10 June 2019).

16. Nugraheni, I.R.; Wijayanti, D.P.; Sugianto, D.N.; Ramdhani, A. Study of inundation events along the southern coast of Java and Bali, Indonesia (case studies 4–9 June 2016). *IOP Conf. Ser. Earth Environ. Sci.* **2017**, *55*, 012014. [CrossRef]

17. Radarcirebon.com Waspada Rob Susulan, Ini Sebabnya—Radarcirebon.com. Available online: http://www.radarcirebon.com/waspada-rob-sususlan-ini-sebabnya.html (accessed on 26 March 2019).

18. Hasan, S.; Rambabu, C. Enhanced Representation of Java Sea Tidal Propagation through Sensitivity Analysis. *J. Water Resour. Hydraul. Eng.* **2017**, *6*, 9–21. [CrossRef]

19. Koropitan, A.F.; Ikeda, M. Three-dimensional modeling of tidal circulation and mixing over the Java Sea. *J. Oceanogr.* **2008**, *64*, 61–80. [CrossRef]

20. Robertson, R.; Ffield, A.; Observatory, L.E.; York, N.; Mountain, C.; Upper, R.; York, N. M 2 Baroclinic Tides in the Indonesian Seas. *Oceanography* **2005**, 1–26. [CrossRef]

21. Khasanah, I.U.; Heliani, L.S.; Basith, A. Trends and Seasonal to Annual Sea Level Variations of North Java Sea Derived from Tide Gauges Data. In Proceedings of the First International Conference on Technology Innovations Society, Padang, West Sumatra, Indonesia, 21 July 2016 – 22 July 2016; pp. 308–316.

22. Andreas, H.; Abidin, H.Z.; Sarsito, D.A. Tidal inundation ("Rob") investigation using time series of high resolution satellite image data and from institu measurements along northern coast of Java (Pantura). In *IOP Conference Series: Earth and Environmental Science*; IOP Publishing: Bristol, UK, 2017; Volume 71, p. 012005.

23. Haigh, I.D. *Tides and Water Levels. Encyclopedia of Maritime and Offshore Engineering*; John Wiley & Sons, Ltd.: Hoboken, NJ, USA, 2017; pp. 1–13. [CrossRef]

24. Marfai, M.A.; King, L. Tidal inundation mapping under enhanced land subsidence in Semarang, Central Java Indonesia. *Nat. Hazards* **2008**, *44*, 93–109. [CrossRef]

25. Wolf, J. Coastal flooding: Impacts of coupled wave–surge–tide models. *Nat. Hazards* **2009**, *49*, 241–260. [CrossRef]

26. Jacobs, J.M.; Cattaneo, L.R.; Sweet, W.; Mansfield, T. Recent and Future Outlooks for Nuisance Flooding Impacts on Roadways on the US East Coast. *Transp. Res. Rec.* **2018**, *2672*, 1–10. [CrossRef]

27. Poulter, B.; Halpin, P.N. Raster modelling of coastal flooding from sea-level rise. *Int. J. Geogr. Inf. Sci.* **2008**, *22*, 167–182. [CrossRef]

28. Van de Sande, B.; Lansen, J.; Hoyng, C. Sensitivity of Coastal Flood Risk Assessments to Digital Elevation Models. *Water* **2012**, *4*, 568–579. [CrossRef]

29. Kumbier, K.; Carvalho, R.; Woodroffe, C. Modelling Hydrodynamic Impacts of Sea-Level Rise on Wave-Dominated Australian Estuaries with Differing Geomorphology. *J. Mar. Sci. Eng.* **2018**, *6*, 66. [CrossRef]

30. Verwaest, T.; Vanneuville, W.; Peeters, P.; Mertens, T.; De Wolf, P. Uncertainty on coastal flood risk calculations, and how to deal with it in coastal management: Case of the Belgian coastal zone. In Proceedings of the 2nd IMA International Conference on flood Risk Assessment, Plymouth, UK, 4–5 September 2007.

31. Budiyono, Y.; Aerts, J.; Brinkman, J.J.; Marfai, M.A.; Ward, P. Flood risk assessment for delta mega-cities: A case study of Jakarta. *Nat. Hazards* **2015**, *75*, 389–413. [CrossRef]

32. Samanta, S.; Koloa, C. Modelling Coastal Flood Hazard Using ArcGIS Spatial Analysis tools and Satellite Image. *Int. J. Sci. Res.* **2014**, *3*, 961–967.

33. Parker, B.B. *Tidal Analysis and Prediction*; NOAA Special Publication NOS CO-OPS 3: Silver Spring, MD, USA, 2007.

34. Seenath, A.; Wilson, M.; Miller, K. Hydrodynamic versus GIS modelling for coastal flood vulnerability assessment: Which is better for guiding coastal management? *Ocean Coast. Manag.* **2016**, *120*, 99–109. [CrossRef]

35. Yang, X.; Grönlund, A.; Tanzilli, S. Predicting Flood Inundation and Risk Using Geographic Information System and Hydrodynamic Model. *Geogr. Inf. Sci.* **2002**, *8*, 48–57. [CrossRef]

36. BPS Cirebon. *Cirebon Regency in Figure*; 32.09.1501; BPS—Statistics of Cirebon Regency: Cirebon, Indonesia, 2016; ISBN 02154242.

37. KKP. *Salt Production of Indonesia*; KKP: Jakarta, Indonesia, 2015; (In Bahasa).

38. PSSDAL. *Peta Lahan Garam Indonesia*; Saputro, G.B., Edrus, I.N., Hartini, S., Poniman, A., Eds.; PSSDAL Bakosurtanal: Cibinong, Indonesia, 2010; ISBN 978-9-79-126669-7.

39. Mustaid, Y.; Tetsuo, Y. Numerical modeling of tidal dynamics in the Java Sea. *Coast. Mar. Sci.* **2013**, *36*, 1–12.

40. Durand, J.R.; Petit, D. The Java Sea Environment. In *Proceedings of the Biodynex: Biology, Dynamics and Exploitation of the Small Pelagig Fishes in the Java Sea*; Nurhakim, S., Ed.; Agency for Agricultural Research and Development: Jakarta, Indonesia, 1995; pp. 15–38.

41. ICCSR. *Scientific Basis: Analysis and Projection of Sea Level Rise and Extreme Weather Event*; ICCSR: Jakarta, Indonesia, 2010.

42. Takagi, H.; Esteban, M.; Mikami, T.; Fujii, D. Projection of coastal floods in 2050 Jakarta. *Urban Clim.* **2016**, *17*, 135–145. [CrossRef]

43. Nagara, G.A.; Sasongko, N.A.; Olakunle, O.J. *Introduction to Java Sea*; University of Stavanger: Stavanger, Norway, 2010.

44. Heidarzadeh, M.; Muhari, A.; Wijanarto, A.B. Insights on the Source of the 28 September 2018 Sulawesi Tsunami, Indonesia Based on Spectral Analyses and Numerical Simulations. *Pure Appl. Geophys.* **2018**, *176*, 25–43. [CrossRef]

45. BIG DEMNAS Seamless Digital Elevation Model (DEM) Dan Batimetri Nasional. Available online: http://tides.big.go.id/DEMNAS/ (accessed on 26 March 2019).

46. Tehrany, M.S.; Pradhan, B.; Mansor, S.; Ahmad, N. Flood susceptibility assessment using GIS-based support vector machine model with different kernel types. *Catena* **2015**, *125*, 91–101. [CrossRef]

47. Zalite, K. *Radar Remote Sensing for Monitoring Forest Floods and Agricultural Grasslands*; University of Tartu Press: Tartu, Estonia, 2016.

48. Arabelos, D.N.; Papazachariou, D.Z.; Contadakis, M.E.; Spatalas, S.D. A new tide model for the Mediterranean Sea based on altimetry and tide gauge assimilation. *Ocean Sci.* **2011**, *7*, 429–444. [CrossRef]

49. Nguyen, T. An Evaluation of Coastal Flooding Risk due to Storm Surge under Future Sea Level Rise Scenarios in Thua Thien Hue Province, Vietnam. Ph.D. Thesis, Texas Tech University, Lubbock, TX, USA, 2017.

50. Thambas, A.H. Inundation Risk Analysis of the Storm Surge and Flood for the Ariake Sea Coastal Disaster Management. Ph.D. Thesis, Saga University, Saga, Japan, 2016.

51. Qiao, H.; Zhang, M.; Jiang, H.; Xu, T.; Zhang, H. Numerical study of hydrodynamic and salinity transport processes in the Pink Beach wetlands of the Liao River estuary, China. *Ocean Sci.* **2018**, *14*, 437–451. [CrossRef]

52. Faber, R. *Flood Risk Analysis: Residual Risks and Uncertainties in an Austrian Context*; University of Natural Resources and Applied Life Sciences: Vienna, Austria, 2006.

53. Reforgiato, D.; Zavarella, V.; Consoli, S. A survey on tidal analysis and forecasting methods for tsunami Detection. *Sci. Tsunami Hazards* **2013**, *33*, 1–58.

54. Pawlowicz, R.; Beardsley, B.; Lentz, S. Classical tidal harmonic analysis including error estimates in MATLAB using TDE. *Comput. Geosci.* **2002**, *28*, 929–937. [CrossRef]

55. Fernández, V.; Silva, R.; Mendoza, E.; Riedel, B. Coastal flood assessment due to extreme events at Ensenada, Baja California, Mexico. *Ocean Coast. Manag.* **2018**, *165*, 319–333. [CrossRef]

56. Kalligeris, N.; Winters, M.; Gallien, T.; Tang, B.-X.; Lucey, J.; Delisle, M.-P. Coastal Flood Modeling Challenges in Defended Urban Backshores. *Geosciences* **2018**, *8*, 450.

57. Department of Environmental Protection. *Rebuild by Design-Hudson River Project Feasibility Study Report*; Department of Environmental Protection: Trenton, NJ, USA, 2014.

58. Elsaesser, B.; Bell, A.K.; Shannon, N.; Robinson, C. Storm surge hind-and forecasting using Mike21FM-Simulation of surges around the Irish Coast. In Proceedings of the DHI International User Conference, Copenhagen, Denmark, 6–8 September 2010; Available online: http//www.dhisoftware.com/GlobalEvents/PastMajorEvents/MIKEByDHI2010/PresentationsAndPapers.aspx (accessed on 17 January 2019).

59. Wang, F.; Hartnack, J.N.; Road, W.S. Simulation of Flood Inundation in Jilin City, Songhua River Project. *City* **2006**.

60. Ningsih, N.S.; Hadi, S.; Utami, M.D.; Rudiawan, A.P. Modeling of storm tide flooding along the southern coast of Java, Indonesia. *Adv. Geosci.* **2011**, *24*, 87–103.

61. Green, J.A.M.; Bowers, D.G.; Byrne, H.A.M. A mechanistic classification of double tides. *Ocean Sci. Discuss.* **2018**, *1*, 1–15. [CrossRef]

62. Seifi, F.; Deng, X.; Andersen, O.B. Accuracy assessment of recent empirical and assimilated tidal models for the Great Barrier Reef region. *Eur. Geosci. Union* **2018**, *20*, 18959.

63. Egbert, G.; Erofeeva, S.Y. Efficient Inverse Modeling of Barotropic Ocean Tides. *Am. Meteorogical Soc.* **2002**, *19*, 183–204. [CrossRef]

64. Taufik, M.; Safitri, D.A. Determining LAT Using Tide Modeling TPXO IO (A Case Study: Fani Island). In Proceedings of the International Conference Data Mining, Civil and Mechanical Engineering, Bali, Indonesia, 4–5 February 2014; Volume 2002, pp. 4–7. [CrossRef]

65. Padman, L. Tide Model Driver (TMD) Manual. 2005, pp. 1–23. Available online: https://svn.oss.deltares.nl/repos/openearthtools/trunk/matlab/applications/DelftDashBoard/utils/tmd/Documentation/README_TMD_vs1.2.pdf (accessed on 14 April 2019).

66. Latief, H.; Putri, M.R.; Hanifah, F.; Afifah, I.N.; Fadli, M.; Ismoyo, D.O. Coastal Hazard Assessment in Northern part of Jakarta. *Procedia Eng.* **2018**, *212*, 1279–1286. [CrossRef]

67. Berlianty, D.; Yanagi, T. Tide and tidal current in the Bali strait, Indonesia. *Mar. Res. Indones.* **2011**, *36*, 25–36. [CrossRef]

68. Gurumoorthi, K.; Venkatachalapathy, R. Hydrodynamic modeling along the southern tip of India: A special emphasis on Kanyakumari coast. *J. Ocean Eng. Sci.* **2017**, *2*, 229–244. [CrossRef]

69. Pugh, D.T. Tides, Surges and mean sea-level. *Mar. Pet. Geol.* **1987**, *5*, 301.

70. Gallien, T.W.; Schubert, J.E.; Sanders, B.F. Predicting tidal flooding of urbanized embayments: A modeling framework and data requirements. *Coast. Eng.* **2011**, *58*, 567–577. [CrossRef]

71. Hanifah, F.; Ningsih, N.S. Identifikasi Tinggi dan Jarak Genangan Daerah Rawan Bencana Rob di Wilayah Pantai Utara Jawa yang Disebabkan Gelombang Badai Pasang dan Variasi Antar Tahunan. *J. Civ. Eng.* **2018**, *25*, 81–86. [CrossRef]

72. Kurniawan, R.; Ramdhani, A.; Sakya, A.E.; Pratama, B.E. High wave and coastal inundation in south of Java and west of Sumatera (Case studies on 7–10 June 2016). *J. Meteorol. Geofis.* **2016**, *17*, 69–76.

73. Châu, V.N. *Assessing the Impacts of Extreme Floods on Agriculture in Vietnam: Quang Nam Case Study*; Massey University: Massey, New Zealand, 2014.

74. Nayak, D.; Bajaji, R. Numerical modeling of tides, currents and waves off Maharashtra coast. *Indian J. Geo-Mar. Sci.* **2016**, *45*, 1255–1263.

75. Kregting, L.; Elsäßer, B. A Hydrodynamic Modelling Framework for Strangford Lough Part 1: Tidal Model. *J. Mar. Sci. Eng.* **2014**, *2*, 46–65. [CrossRef]

76. Webster, T.L. Flood risk mapping using LiDAR for annapolis Royal, Nova Scotia, Canada. *Remote Sens.* **2010**, *2*, 2060–2082. [CrossRef]

77. Simpson, A.L.; Balog, S.; Moller, D.K.; Strauss, B.H.; Saito, K. An urgent case for higher resolution digital elevation models in the world's poorest and most vulnerable countries. *Front. Earth Sci.* **2015**, *3*, 50. [CrossRef]
78. Raji, O.; Del Rio, L.; Gracia, F.J.; Benavente, J. The use of LIDAR data for mapping coastal flooding hazard related to storms in Cadiz Bay (SW Spain). *J. Coast. Res.* **2011**, 1881–1885. Available online: https://www.jstor.org/stable/26482503?seq=1#metadata_info_tab_contents (accessed on 27 April 2019).
79. Chen, W.B.; Liu, W.C. Modeling flood inundation induced by river flow and storm surges over a river basin. *Water (Switzerland)* **2014**, *6*, 3182–3199. [CrossRef]
80. Rey, W.; Salles, P.; Mendoza, E.T.; Torres-Freyermuth, A.; Appendini, C.M. Assessment of coastal flooding and associated hydrodynamic processes on the south-eastern coast of Mexico, during Central American cold surge events. *Nat. Hazards Earth Syst. Sci.* **2018**, *18*, 1681–1701. [CrossRef]
81. Gallien, T.W. Validated coastal flood modeling at Imperial Beach, California: Comparing total water level, empirical and numerical overtopping methodologies. *Coast. Eng.* **2016**, *111*, 95–104. [CrossRef]

International Journal of
Geo-Information

Review

UAV-Based Structural Damage Mapping: A Review

Norman Kerle [1,*], **Francesco Nex** [1], **Markus Gerke** [2], **Diogo Duarte** [3,4] and **Anand Vetrivel** [5]

1. Faculty of Geo-Information Science and Earth Observation (ITC), University of Twente, 7500 AE Enschede, The Netherlands; f.nex@utwente.nl
2. Technische Universität Braunschweig, Institut für Geodäsie und Photogrammetrie, Bienroder Weg 81, 38106 Braunschweig, Germany; m.gerke@tu-bs.de
3. Department of Mathematics, University of Coimbra, Apartado 3008 EC Santa Cruz, 3001-501 Coimbra, Portugal; diogoavaduarte@mat.uc.pt
4. Institute for Systems Engineering and Computers, University of Coimbra, Rua Sílvio Lima, Pólo II, 3030-290 Coimbra, Portugal
5. Experian Singapore Pte. Ltd., 10 Kallang Ave #14-18 Aperia Tower 2, Singapore 339510, Singapore; anand.vetrivel@experian.com
* Correspondence: n.kerle@utwente.nl; Tel.: +31-53-4874-476

Received: 22 November 2019; Accepted: 23 December 2019; Published: 26 December 2019

Abstract: Structural disaster damage detection and characterization is one of the oldest remote sensing challenges, and the utility of virtually every type of active and passive sensor deployed on various air- and spaceborne platforms has been assessed. The proliferation and growing sophistication of unmanned aerial vehicles (UAVs) in recent years has opened up many new opportunities for damage mapping, due to the high spatial resolution, the resulting stereo images and derivatives, and the flexibility of the platform. This study provides a comprehensive review of how UAV-based damage mapping has evolved from providing simple descriptive overviews of a disaster science, to more sophisticated texture and segmentation-based approaches, and finally to studies using advanced deep learning approaches, as well as multi-temporal and multi-perspective imagery to provide comprehensive damage descriptions. The paper further reviews studies on the utility of the developed mapping strategies and image processing pipelines for first responders, focusing especially on outcomes of two recent European research projects, RECONASS (Reconstruction and Recovery Planning: Rapid and Continuously Updated Construction Damage, and Related Needs Assessment) and INACHUS (Technological and Methodological Solutions for Integrated Wide Area Situation Awareness and Survivor Localization to Support Search and Rescue Teams). Finally, recent and emerging developments are reviewed, such as recent improvements in machine learning, increasing mapping autonomy, damage mapping in interior, GPS-denied environments, the utility of UAVs for infrastructure mapping and maintenance, as well as the emergence of UAVs with robotic abilities.

Keywords: drone; computer vision; point clouds; machine learning; CNN; GAN; first responder; RECONASS; INACHUS

---

1. Introduction

1.1. Structural Damage Mapping with Remote Sensing

The first documented systematic post-disaster damage assessment attempt with remote sensing technology dates back to 1906, when parts of earthquake-affected San Francisco were mapped with a 20 kg camera that was raised on a series of kites some 800 m above the disaster scene [1]. This makes damage mapping one of the oldest applications in the remote sensing domain, but also one of the few that continues to elude robust operational solutions, and which remains a subject of active research. Since the early pioneering days, nearly every type of active and passive sensor has been mounted on airborne platforms that range from tethered to autonomous or piloted, as well as satellites operating in different orbital or network configurations, to attempt increasingly automated damage detection [2,3]. However, despite more than a century of research and tremendous technological developments both on the hardware and the computing side, operational image-based damage mapping, such as through the International Charter "Space and Major Disasters" or the Copernicus Emergency Management Service (EMS), continues to be a largely manual exercise (e.g., [4,5]).

Charter and EMS activations center on a particularly challenging type of damage mapping. Both need to respond to a wide range of natural and anthropogenic disaster types, and the first maps are expected to be available within hours of image acquisition, while the particular damage patterns and their recognition are subject to a number of variables. Building typologies, spatial configurations, and construction materials differ, and recognizable damage indicators are strongly dependent on the type of hazard and its magnitude. Image type, in terms of spatial and spectral characteristics, as well as incident angle, but also environmental conditions such as haze or cloud cover, differ enormously, further challenging the development of generic and widely applicable damage detection algorithms. Satellite-based damage mapping has the additional disadvantage that damage that may be quite variably expressed on each of the building's facades, its roof, as well as its interior, is largely reduced to a single dimension, the quasi-vertical perspective that centers on the roof. Damage detection in reality is then supported by the use of proxies, such as evidence of nearby debris or damage clues associated with particular shadow signatures [6,7]. There have been some notable successes in satellite-based damage mapping, especially related to cases where radar data have an advantage, in particular interferometric [8] and polarimetric synthetic aperture radar [9]. Where damage patterns are structurally characteristic, such as foundation walls remaining after the 2011 Tohoku (Japan) tsunami, simple backscatter intensity has also been used to detect damage [10]. Increasingly advanced machine learning algorithms, including convolutional neural networks (CNN), are used to detect different forms of building damage with radar data [11].

Efforts to process optical satellite data for rapid damage mapping are also moving in the machine learning direction. This includes methods based on artificial neural networks [12], and increasingly also CNN [13–16]. Studies vary in terms of mapping ambition, with many only aiming at a binary classification (damage/no damage; [12]), and there is no evidence yet of emerging methodologies being used operationally. However, the recently released xBD satellite dataset containing more than 700,000 building damage labels and corresponding to 8 different disaster types [17] will help in developing and benchmarking novel methodologies.

1.2. Scope of the Review

Automated satellite-based damage mapping has thus shown limited progress, at least in terms of versatile methodologies that can readily map structural damage caused by different event types in diverse environments. At the same time, the proliferation and rapidly growing maturity of unmanned aerial vehicles (UAVs/drones) in recent years has created vast new prospects for rapid and detailed structural damage assessment, which are the focus of this review. We do not consider historical, mainly military systems, such as unpiloted reconnaissance aircraft that date back to World War II. Rather, we focus on the suite of platforms that evolved from remote-controlled (mainly hobbyist) planes and helicopters, with the first documented scientific studies on UAV-based disaster response dating back to about 2005 [18]. The review also does not include non-structural damage assessment, such as studies on crop or forest damage. It also does not cover issues of UAV communication (e.g., use of UAVs to create ad hoc communication networks over disaster areas), nor studies on drone network or scheduling optimization. For both good reviews already exist (e.g., [19,20]).

The review includes peer-reviewed publications indexed in Scopus and Web of Science, focusing on research on automated damage detection rather than provision of data for visual assessment, and is not meant to be exhaustive. While the topic is a niche within the remote sensing domain and the amount of studies remains relatively small, a number of application papers without significant novelty exist, which are excluded here. The article is built on a recent conference contribution [21], though the focus of that paper on the results of two European research projects is expanded here to a comprehensive review study. In addition to tracing relevant technical and methodological developments in damage detection, we synthesize the current state of the art and evaluate current and emerging research directions. In addition, we assess the actual usability and practical value of emerging methods for operational damage mapping, including for local mapping by first responders. In the following section relevant publications on the use of UAVs for structural damage mapping are reviewed, sorted by increasing technical sophistication, and a summary is provided in Table 1.

2. UAV-Based Damage Mapping

2.1. Scene Reconnaissance and Simple Imaging

The principal advantage of a UAV in a disaster situation is its vantage point, a flexible position that can provide both synoptic and detailed views of a potentially complex scene, as well as overcome access limitations. Early studies thus focused on scene imaging, aiding disaster responders by supplying a relatively low-cost aerial perspective [22]. Taking advantage of increasingly efficient structure from motion (SfM) and 3D reconstruction concepts emerging at the time (e.g., [23]), in some early studies data were already processed to derive georeferenced images [24], terrain information/digital elevation models (DEM) [18], or orthophotos [25]/orthomosaics [26]. In cases without a full processing pipeline and where no suitable DEM data existed, pseudo-orthorectified images (assuming constant terrain height) were created. Instead of still images also video data were transmitted in real time to allow visual damage inspection [27].

In the years following the initial studies, little methodological progress in damage mapping was made, despite advances in off-the-shelf UAV systems, or the emergence of ArduPilot in 2007 for improved UAV flight stability, or Pix4D(Pix4D, Switzerland) in 2011 for easier photogrammetric image processing. A range of studies appeared that essentially still focused on image provision or simple photogrammetric processing, using remote-controlled helicopter systems [28], multi-copters [29–32], or fixed-wing UAVs [33–35].

2.2. Texture- and Segmentation-Based Methods

Initial attempts to extract damage information automatically from UAV data were based on segmentation- and texture-based approaches, using mono-temporal imagery. Fernandez Galarreta et al. [36] processed UAV imagery of an 2012 Emilia Romagna (Italy) earthquake site into detailed 3D models. The work adapted and expanded earlier approaches developed for the airborne (piloted) Pictometry system that yields similar oblique, overlapping, and multi-perspective imagery. Also, those images had been photogrammetrically processed [37], and used for structural damage assessment [38,39]. The analysis of [36] focused on geometric damage indicators such as slanted walls or deformed roofs, as well as presence of debris piles (Figure 1). In addition, object-based image analysis (OBIA) was carried out on the images to extract damage features such as cracks or holes, but also identification of those damage features intersecting with apparent load-carrying structural elements. A similar OBIA strategy was used by [40] to identify damage in Mianzhu city, affected by the 2008 Wenchuan earthquake.

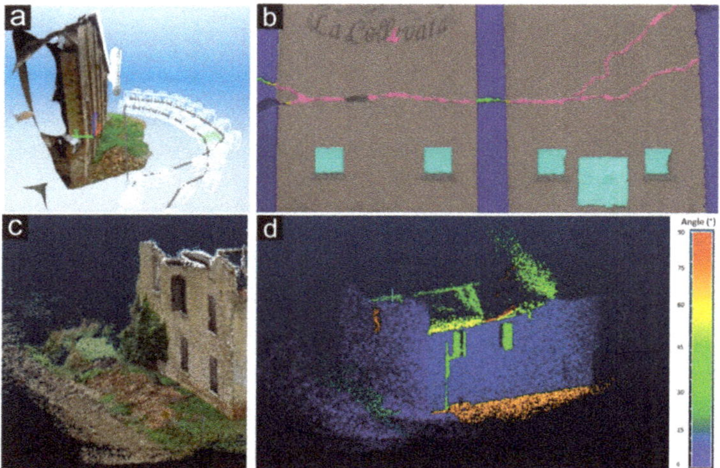

Figure 1. Damages identified from unmanned aerial vehicle (UAV)-derived point clouds and from object-based image analysis (OBIA) processing. (**a**) Inclination in walls, (**b**) openings (turquois), cracks (magenta), and damage crossing beams, (**c**,**d**) detailed point cloud and segment orientation angles [adapted from 36].

2.3. Conventional Classifiers

The work of Fernandez Galarreta et al. demonstrated the significance of geometric information in damage detection, in particular of openings in roofs and façades. Vetrivel et al. [41] advanced the work by developing a method to isolate individual buildings from a detailed image-derived point cloud covering a neighborhood of Mirabello (Italy) comprising nearly 100 buildings. Each of those was then subjected to a search for openings attributable to seismic damage, such as partial roof collapses or holes in the façades, a focus similar to [42]. The gaps were identified based on Gabor wavelets as well as histogram of gradient (HoG) orientation features. Two basic machine learning algorithms, Support Vector Machine (SVM) and Random Forest (RF), were used to identify damaged regions based on the radiometric descriptors, with a success rate of approximately 95%. However, the work also illustrated how the segmentation of point clouds is frequently hindered by artefacts and data gaps. In [43], an approach was developed to overcome this problem: after projecting the initial point cloud-derived 3D segments into image space, a subsequent segmentation using both geometric and radiometric features yielded more accurate and complete building segments.

Table 1. Summary the relevant technical papers reviewed in Section 2, organized by level of technical sophistication.

Processing Level	Publications	Platform	Notes
Scene reconnaissance/simple imaging	Whang et al., 2007 [22] Adams et al., 2014 [29] Dominici et al., 2012 [30] Mavroulis et al., 2019 [31]	Multi-copter	• Studies focusing on visual image analysis, simple 3D terrain reconstruction, as well as creation of orthophotos or orthomosaics
	Nakanishi and Inoue, 2005 [18] Murphy et al., 2008 [27] Kochersberger et al., 2014 [28]	Helicopter	
	Lewis, 2007 [26] Suzuki et al., 2008 [24] Bendea et al., 2008 [25] Hein et al., 2019 [33] Xu et al., 2014 [34] Gowravaram et al., 2018 [35]	Fixed-wing	• Includes platforms with specialized abilities, such as deployment of spectrometers for gas detection [28]
Texture- and segmentation-based methods; change detection	Fernandez Galarreta et al., 2015 [36] Dorafshan et al., 2018 [44] Chen et al., 2019 [45] Akbar et al., 2019 [46]	Multi-copter	• Includes approaches based on handcrafted texture features (e.g., Gabor wavelets, histogram of gradient),
	Zeng et al., 2013 [40]	Helicopter	
	Kakooei and Baleghi, 2017 [47]	Fixed-wing	• Change detection based on 3D features only (e.g., [48])
	Grenzdorffer et al., 2008 [37]		
	Gerke and Kerle, 2011a [38] Gerke and Kerle, 2011b [39] Zeng et al., 2013 [40] Vetrivel et al. 2016 [47] Tu et al., 2017 [49] [1]	Pictometry [2]	• Fusion of pre- and post-event satellite imagery with both manned airborne and UAV data [47]
Conventional classifiers	Li et al., 2015 [42] Vetrivel et al., 2015b [43]	Multi-copter	• Includes methods based on classifiers such as Support Vector Machine (SVM) and Random Forest, as well as image fusion
	Vetrivel et al., 2015a [41]	Pictometry	• Combination of image and 3D structural features
	Lucks et al., 2019 [50]	– [3]	• Use of boosting (e.g., AdaBoost)

Table 1. *Cont.*

Processing Level	Publications	Platform	Notes
Advanced machine learning/CNN/generative adversarial networks (GAN)	Duarte et al., 2017 [51] Dorafshan et al., 2018a [52] Dorafshan et al., 2018b [53] Xu et al., 2018 [54] Vetrivel et al., 2018 [55]	Multi-copter	• Use of active learning methods, convolutional neural networks (CNN), as well as convolutional autoencoders (CAE)
	Cusicanqui et al., 2018		• Use of boosting (e.g., XGBoost)
	Duarte et al., 2018a [56]		• Combination with
	Duarte et al., 2018b [57]		• Conventional classifier
	Kerle et al., 2019 [58]		• Use of single-shot multi-box detector algorithm
	Nex et al., 2019a [59]		• Use of UAV data to classify satellite imagery [56]
	Tsai and Wei, 2019 [60]		• Multi-resolution CNN
	Xu et al., 2018 [54]	Fixed-wing	• Semantic segmentation
	Vetrivel et al., 2018 [55]	Pictometry	• Transfer learning
	Duarte et al., in press [61]		• Morphological filtering
	Li et al., 2018 [62] Li et al., 2019 [63]	$_-2$	• Bayesian optimization

Table 1. *Cont.*

Processing Level	Publications	Platform	Notes
	Liang et al., 2019 [64]	Ground-based	• Emerging use of generative adversarial networks (GAN) [60]
	Nex et al., 2019b [65]	Multiple platforms	• Review of the performance of state-of-the-art CNN
	Song et al., 2019 [66] Huang et al., 2019 [67] Duarte et al., 2018 [56]	Manned airborne	• Includes combination of deep learning and SLIC superpixels, as well as multi-scale analysis

[1] Manned system with 5 cameras comparable to Pictometry. [2] Though not a UAV category, some relevant Pictometry-based studies are included in the review. [3] Use of generic high-resolution airborne data, with no specific platform being indicated.

The work in [41] work also showed the limitations of HoG and Gabor filters in the classification of complex scenes, and of global feature representations on general. The latter cause problems when scene and image characteristics vary, which is typically the case between different disaster areas or in multi-temporal assessments. The work described in [68] moved towards descriptors that are more generalizable and invariant to image characteristics. The method was built on the Visual Bag of Words approach and focused on the detection of rubble, debris piles, and severe spalling. The method performed well on individual UAVs and also Pictometry data sets of Mirabello (Italy) and Port-au-Prince (Haiti), respectively, but also on a dataset that combined the two airborne datasets with transverse street-level images. The limitation of the method is that it is grid-based and can only identify general damage patches, i.e., grid cells affected by one or more of the damage types considered, a limitation also evident in the study of [50], who used RF on superpixels. A detailed localization and characterization (size, shape, etc.) of damages of a specific type would be preferable, though this will come at the cost of increased processing time.

2.4. Advanced Machine Learning and the Emergence of CNN

Image classification used for damage mapping increasingly made use of machine learning, in particular SVM and RF [41,54,69] or different boosting algorithms, such as AdaBoost [38] or XGBoost [58], and moving towards more advanced scene understanding and semantic processing. However, the features used were typically hand-crafted (such as HoG or Gabor, or other point feature descriptors related to spectral, textural, and geometrical properties [54]), and emerging work had shown that in deep learning approaches CNN could actually learn features and their representation directly from the image pixel values [70]. Thus, the damage detection work proceeded in this direction, hypothesizing that image classification would benefit from the micropropagation of 3D point cloud features. The work described in [55] applied a multiple-kernel-learning framework on several sets of diverse aerial images, and showed that combining the radiometric and geometric information yields higher classification accuracies. The processing was based on Simple Linear Iterative Clustering (SLIC) superpixels, meaning that damage was again only identified in patches, though those were labelled with specific prediction scores. Song et al. [66] also worked with SLIC superpixels, though unlike in [55] where they had formed the basis for the ML analysis, here first a CNN-based damage detection was carried out directly on the image, and the SLIC segments were then used in combination with mathematical morphology to refine the results. In [67] a similar approach was taken, except that instead of SLIC a multi-resolution segmentation was carried out, to allow features naturally occurring at different spatial scales to be used effectively. The CNN approach developed in [55] was also used by Cusicanqui et al. [71], who reasoned that video data are often available before suitable still photographs (e.g., acquired by police or the media). In the study it was thus tested whether 3D reconstructions based on video data could offer similar support, and it was indeed shown that a binary damage classification based on deep learning applied to SLIC superpixels and the 3D models led to results comparable to those based on still photographs.

The particular significance of the work in [55] for disaster response and search and rescue was that the method demonstrated significant transferability, which has become a frequent focus in recent literature. A model trained with a sufficient number of samples (e.g., trained before an actual event) performed well when then applied to a new disaster scene, supporting a rapid analysis without the need for extensive retraining. This approach can help to overcome the traditional limitation of CNN, i.e., their need for a large amount of labelled training data. A different approach was taken by Li et al. [62], who used a convolutional autoencoder (CAE) that was trained using unlabelled post-disaster imagery based on SLIC superpixels, with results being finetuned by a CNN classifier. In follow-up work [63] the authors in addition employed a range of data augmentation methods, such as data blurring or rotating, to enlarge the number of samples. The resulting pre-training improved the overall damage detection accuracy by 10%.

Disaster scenarios are frequently characterized by imperfect image data availability, and a rapid response effort has to make do with what exists. In this respect it is valuable to be able to incorporate images of different types and scales into the training model. Duarte et al. [56] trained a CNN with different types of aerial imagery to classify post-disaster satellite data of Port-au-Prince. Although information coming from the different image resolutions evidently improved the model and classification accuracy, the approach still failed to capture smaller damage features. The work also focused on determining the effect of multi-scale information on the CNN activation layers as a proxy for improved damage recognition, while not allowing a detailed assessment of where the classification improvement originated in terms of false positives and negatives, or specific damage types. Later work focused on multi-resolution feature fusion and its effect on building damage classification [57]. It showed that such a fusion is useful and can improve the overall accuracy, though it still failed to show which specific damage types are identified, and how well they are captured.

Earlier work had shown how highly variable the expression of structural damage is in vertical and oblique data [6]. The former essentially only considers the damage expressed in the roof, and in addition makes use of proxies such as debris piles for specific shadow configurations [7]. Significant additional information is also encoded in the façade information, as already explained in Section 2.1. However, the OBIA-based approach used for example in [36] tends towards overfitting and lacks the efficiency and transferability of deep learning. While a focus on façades is appealing, their actual delineation in imagery poses its own challenges, especially when considering aspects such as occlusion or environmental effects such as shadows (Figure 2). The work described in [51] thus focused on developing an efficient method to extract façades that were subsequently assessed for damage using CNN. The approach made use of a point cloud calculated from vertical imagery acquired in an initial UAV survey. From the sparse point cloud, the building roofs were segmented and the building façades hypothesized, which in turn was used to extract the actual façades from oblique UAV images. The patch-based damage classification had an overall accuracy of approximately 80%, though the work also demonstrated the significant challenge of damage identification on façades, due to architectural complexities and associated diverse shadow patterns, but also occlusion (by external features such as vegetation, or internal ones such as balconies).

Figure 2. Typical problems for image processing posed by shadow and occlusion [51].

It stands to reason that some ambiguities can be resolved by analysing multi-perspective data (views of a given façade from different angles that go beyond regular stereoscopic overlap), but also by incorporating multi-temporal data where available. The majority of the studies described above only used post-disaster imagery. However, in the last few years the availability of high spatial resolution pre-event reference imagery has been growing rapidly. This has led to additional methodological developments that built on the segmentation- and texture-based damage detection described above, extending them into a multi-temporal framework. Vetrivel et al. [47] used pre- and post-earthquake data of L'Aquila (Italy) and focused on the identification of 3D segments missing in the post-disaster data as an indicator of damage. Both voxel- and segment-based approaches were tested, and finally a composite segmentation method that subjects an integrated pre- and post-event point cloud to plane-based segmentation was chosen. Although working with conventional airborne data, in [61] those assumptions were also tested in a CNN framework, where 6 different multi-temporal approaches were compared against 3 mono-temporal ones. It was concluded that a multi-temporal approach with 3 views at each the pre- and post-event epoch performed best. Also, here smaller damage features eluded detection. However, the authors expect better results with UAV data, given that the problem of occlusion can be reduced through more flexible image acquisition.

2.5. Levels of Disaster Damage Mapping

Early efforts in disaster response with satellite imagery identified damaged areas more generally, while airborne data were used to detect specific damage proxies, usually debris piles (e.g., [72]). Especially in more recent years, overall classification accuracy and f-scores have been the most commonly used metrics to assess the efficacy of a given damage mapping method, and to judge progress within the discipline. However, this focus neglects an inherent incomparability of many of the studies produced to date, and the absence of a generally agreed upon damage scale. The introduction of the European Macroseismic Scale 1998 (EMS-98) led to a broad homogenization and alignment of efforts, by grouping structural building damage in 5 categories, D1 (negligible/slight damage)–D5 (destruction) [73]. Building on its common use in satellite-based damage detection (e.g., [74–76]), later its utility for UAV-based damage mapping was explored. For example, [38] classified building damage according to EMS-98, though recognizing the diversity and ambiguity of the observed damage patterns the study did not aim at automatic damage classification, except in cases where the 3D model clearly showed complete collapse (D5). Also, studies [36,48,77] used this scale as a basis, with [31] even adding a 6th damage level.

One consequence of the continuing challenge of image-based damage mapping is that, while D1 and D5 are comparatively easy to determine but intermediate damage stages are not, many studies have departed from the 5-level classification scheme. The work in [50] opted instead for a 4-class approach (intact, light, medium, and heavy damage), while several studies grouped damage into 3 classes. However, even within one such category damage levels/class names vary, limiting comparability. For example, Zeng et al. [40] mapped intact, damaged, and destroyed buildings, while Vetrivel et al. [47] termed the classes undamaged, lower levels of damage, and highly damaged/collapsed, and Song et al. [66] distinguished intact, semi-collapsed, and collapsed buildings, with differences in class definition going beyond semantics. However, the majority of recent studies opted for a simple binary classification, either explicitly mapping both damaged and undamaged structures (e.g., [12,49,55,65,67]), or only mapping damage in general in a single class [51,61]. In addition, there are studies that focused on the identification of specific damage types, such as holes in the roof [41,42], or dislocated roof tiles and cracks along walls [36]. Others mixed damage and proxy classes, such as [62], who mapped damaged and undamaged structures, but also debris as a separate class. Creative choice of class names is further hindering a comparison between different studies. Li et al. [63] used the classes mildly damaged and ruins, while Xu et al. [54] mapped categories including roof, ground, debris, and small objects. The difficulty of image-based damage mapping has led to a focus on severe damage classes (D4-5), making studies such as [42] that expressly focus on lesser damage (D2-3) an exception. Approaches based on deep learning are particularly suited for binary classification, which is another reason why in the interest of automation only a single damage class is now frequently considered.

2.6. The Special Case of Infrastructure Damage Mapping

The focus of this review is on structural building damage. However, one of the fastest growing UAV application areas in recent years is infrastructure monitoring and detection of damage indicators related to wear and degradation, such as of roads, bridges, or tunnels. The lines between disciplines have blurred, with studies such as by Dominici et al. [32] addressing both regular structures and infrastructure. Furthermore, from a methodological perspective studies focusing on crack or spalling assessment along bridges or tunnels are also relevant for the disaster damage mapping community, and damage to infrastructure caused by disaster events naturally also falls under the scope of this review. For this reason, papers marking key developments in infrastructure monitoring and damage mapping are briefly reviewed here.

A recent review by Dorafshan and Maguire [52] provides an overview of the specific challenges of bridge inspection and maintenance, and how UAVs, both with active and passive sensors, are starting to become a commonly used tool. In an early study by Whang et al. [22], a UAV with two coaxial rotors was developed to perform somewhat autonomous bridge inspection, within limits even in GPS-denied areas beneath the bridge. In addition, the system was able to place a small autonomous rover on the bridge using ultrasonic localization, and which provided images for damage inspection. However, few details about the actual methods and system performance are provided in the paper. The authors of [44] focused on the detection of small fatigue cracks on bridges, assessing the value of active illumination, and carrying out controlled laboratory experiments to determine detection limits and optimal mapping approaches.

Increasingly, the focus has been on image- or laser-based 3D reconstruction of the bridge or tunnel in question, as a basis for visual or automated damage identification. In [78] the accuracy and thus utility of such 3D models was assessed, and [45] also assessed how well complex bridge structures can be reconstructed with SfM methods, in addition attempting 3D volume calculations or major spalling instances. The work of [79] expressly focused on seismic damage detection on bridges, also using UAV-based 3D reconstructions, though here starting with pre-event Building Information Modelling (BIM) data that were updated with the detected damage. Akbar et al. [46] addressed structural health monitoring (SHM) of tall structures, focusing on comprehensive 3D model creation through speeded up robust features (SURF), and on the detection of simulated damage features on large concrete slabs, though providing little detail on the actual damage detection algorithm.

Deep learning with CNN is also being used in SHM. In [53] an AlexNet network was trained to detect small cracks in concrete walls, reporting accuracies of nearly 95%, and also testing network transferability. Comparable accuracies were reported by Liang [64], who in addition also tested GoogleNet and VGG-16 networks to detect earthquake damage on a bridge.

3. Damage Product and System Usability

Post-disaster damage mapping serves a specific purpose, i.e., providing timely, accurate, and actionable information to a range of stakeholders. Those include civil protection agencies planning emergency response actions, but also incident commanders and first responders operating at the actual disaster site. One of the consequences of the growing availability of UAV technology is a declining need to rely on formal protocols such as the Charter or EMS, and instead allowing actual site-based damage mapping. It is thus surprising that the usability of data acquisition pipelines (including planning tools, hardware components, and data processing routines), but also of resulting damage mapping products, has scarcely been considered in the literature reviewed in this paper. This section briefly introduces two recent research projects with a strong focus on UAV-based structural damage assessment, and from which a number of publications reviewed in this paper emerged. In these projects also a range of different end users participated, and their evaluation of the developed damage mapping procedures is also summarized.

3.1. Damage Detection in Two European Research Projects

RECONASS (Reconstruction and Recovery Planning: Rapid and Continuously Updated Construction Damage, and Related Needs Assessment; www.reconass.eu) and INACHUS (Technological and Methodological Solutions for Integrated Wide Area Situation Awareness and Survivor Localization to Support Search and Rescue Teams; www.inachus.eu) were research projects funded through the 7th Framework of the European Union, and which ran with some overlap from 2013 until the end of 2018. The focus of RECONASS was to create a system for monitoring and damage assessment for individual high-value buildings, based on a range of internally installed sensors that included accelerometers, inclinometers, and position tags, with data getting processed in a finite element structural stability model to determine damages caused by seismic activity or by either interior or exterior explosions. UAV-based 3D reconstruction of the building exterior and detailed damage mapping were carried out to patch data gaps caused by failed sensor nodes, as well as to validate model outputs. The progressively developed methods were tested in a series of experiments, culminating in a pilot where a 3-story reinforced concrete building was first subjected to an explosion of 400 kg TNT placed 13 m away, and later by a 15 kg charge detonated within the structure itself. End users, including the German Federal Agency for Technical Relief (THW), were present to assess the utility of the system.

The purpose of INACHUS was to assist disaster response and urban search and rescue forces by providing early and increasingly detailed information on damage hotspots and the likely location of survivors. Different UAV platforms, but also ground-based and portable laser scanning instruments, were used to map a damaged structure. One research focus was on scene reconstruction and damage mapping based on optical imagery from a low-cost UAV. The French remote sensing lab ONERA also deployed various larger UAVs that carried different laser scanners, in part with proprietary data processing solutions. The major pilots were also assessed by a group of end users.

3.2. Tests with End Users in Two European Research Projects

Both RECONASS and INACHUS included a number of pilot experiments, where first individual components or sets thereof, and later the entire systems were tested under relatively realistic conditions. For the explosion experiments in Sweden data were acquired using an Aibot X6 Hexacopter carrying a Canon D600 camera with a Voigtländer 20 mm lens. In addition to reference data, images were acquired after both the exterior and the interior blasts, with a ground sampling distance (GSD) of approximately 1.5 cm. From those images, detailed 3D point clouds were calculated and analyzed. The data proved suitable to identify damage-related openings, such as infill walls damaged or blown-out by the blasts, as well as cracks and debris. Additionally, subtle façade deformations could be detected and quantified (Figure 3), both using only the post-detonation point cloud, as well as in a comparison with pre-event reference data. It was also shown how a BIM model of the structure could be automatically updated, both to visualize and catalogue detailed damage information. THW deployed a LEICA TM30 total station to survey the structure from 4 reference points, using 16 prisms mounted on the structure. While the total station has the advantage that a structure can be continuously monitored for minute deformations—critical when rescue personnel operates near or within weakened structured—the UAV-derived data provided damage data of comparable quality, with greater flexibility and lower cost, including the roof that ground-based surveys cannot see, and potentially operated from a safer distance. The building was further surveyed by a Riegl VZ400 terrestrial laser scanner (TLS), which also confirmed the high quality of the UAV-derived 3D models.

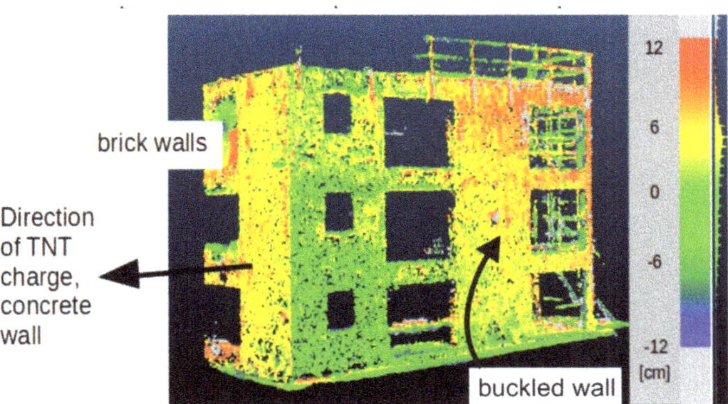

Figure 3. UAV-derived point clouds of reinforced concrete structure with brick in-fill walls subjected to exterior and interior detonations. Openings, cracks, and debris piles, as well as subtle deformation in the façades were automatically detected.

Four INACHUS pilot experiments were conducted at 4 different sites in France and Germany, and included buildings in the process of being demolished, as well as an urban search and rescue training site (Training Base Weeze in Germany). In response to criticism by end users in RECONASS as to the high cost of the Aibot UAV (ca. 40,000 Euro), in INACHUS low-cost DJI drones (Phantom 4 and Mavic Pro) were used. Following the research directions described in Sections 2.3 and 2.4, the work focused less on simple scene reconstruction, but on integration with other spatial data, as well as advanced data analysis, including with CNN. For each of the pilots, the building in question was also surveyed by ONERA using different UAV-borne laser instruments, as well as with a TLS, to detect the respective strengths of the individual systems. The initial experiments with UAV-based laser scanners failed. First a Riegl VZ-1000 instrument (weight of about 10 kg) was deployed on a Yamaha RMAX helicopter (weight > 60 kg), though the acquired data suffered from artefacts and were not useful. Also, data acquired with a Velodyne HDL32 (weight of only 1.3 kg) deployed on a VARIO BENZIN helicopter (weight just under 10 kg) proved unusable for damage detection, owing to the very unstable platform. For the final pilot, a high quality Riegl VUX-1 was mounted on a stable DJI Matrice 600 hexacopter platform. The data were excellent, though the combined system is also very costly (>80,000 EUR) and requires expert knowledge for flight planning and execution, as well as data processing. The mapping with optical data focused on using data acquired with the built-in cameras of the Phantom 4 and Mavic Pro (costs of < 2000 Euro), and advanced along the computer vision and machine learning trajectory described earlier. The 3D data obtained from the optical imagery were of comparable quality to the VUX-1 data while also providing native color information, better spatial detail, and full coverage also of façades (Figure 4). The expectation that the airborne laser data would patch the one principal weakness of photogrammetry, the inability to map dark interior spaces through openings (as a means of possibly locating trapped survivors), was also not met. The data on openings and connected interior spaces were primarily delivered from the tripod-mounted ground-based laser scanner, though here the limited flexibility and occlusion by the building's structural elements also prevented a complete mapping of openings.

While commercial UAVs by DJI and other makers have clearly reached high levels of cost-benefit, stability, and reliability, most are also not designed to be survey-grade instruments working in real time. For rapid search and rescue support it is vital to provide usable information quickly. For that reason, in INACHUS a procedure was developed to process the data with minimal delay. Working with the ability of the Mavic Pro to stream images during flight, a procedure was built that (i) downloads images right after acquisition, (ii) builds a progressively extended sparse 3D model of the scene using established SfM methods, (iii) applies CNN to detect damage, and (iv) orthorectifies the images using the 3D model. By the time the UAV lands after a maximum flight duration of about 25 min, all processing is done and the damage map available. A smart phone app was also built that allows this procedure to be executed together with a standard laptop (Figure 5). Details about the app and data processing workflow can be found in [59], while more information about the optimized CNN that was made available on GitHub can be found in.

Figure 4. Point cloud representation of an INACHUS pilot structure in Lyon, France, calculated from optical imagery acquired with a low-cost commercial drone (Phantom 4, DJI), showing damage detected through machine learning (red).

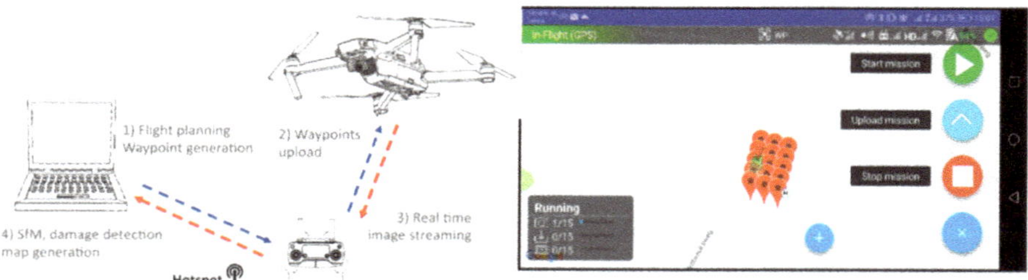

Figure 5. Workflow of the app developed for near real-time damage mapping. Images are streamed to a laptop computer and processed immediately after acquisition. A convolutional neural network (CNN)-based damage detection algorithm is applied, and a progressively built sparse 3D model is used to orthorectify them. By the time the UAV lands, an orthomosaic displaying the damage is finished (adapted from [59]).

3.3. Validation

At every pilot, different end users were present and undertook a detailed assessment of every tool produced and tested. The RECONASS system was evaluated by THW at the pilot site, and more extensively in a dedicated workshop at ISCRAM 2017 by a total of 11 specialist end users, representing both governmental and non-governmental emergency response organizations, as well as organizations involved in the creation of damage maps. It was concluded that the UAV-based element met all previously established user requirements, principally the detection of all externally expressed damage types and their annotation both on imagery but also a 3D model and a BIM, as well as the provision of 3D volume calculations, all in GIS-ready format. The final system received a maximum score of 10/10.

At the final INACHUS pilot that took place in Roquebillière, France, in November 2018 a total of 25 end users from 8 countries participated, representing USAR teams and other civil protection organizations. They followed individual demonstrations of all technical tools developed and graded them. Of all hard- and software or procedure solutions developed in INACHUS, the 3D mapping and damage detection with a light-weight commercial UAV scored highest (overall 4.5 out of 5). The high score does not so much represent a high level of technical sophistication, but rather the simplicity, both in terms of off-the-shelf hardware and an automated flight planning and damage mapping routine. The end users especially appreciated the simple, low-cost approach that provided accurate and useful information in near-real time, without the need for a highly specialized operator.

3.4. Limitations

Despite the positive evaluations, the end user assessment also revealed limitations of the developed damage mapping solution. Legal restrictions of drone deployment continue to pose challenges, though problems are less severe for lighter platforms, and in addition first responder and civil protection organizations tend to operate under different legal frameworks. A clear disadvantage of small multi-copter UAV platforms is their comparatively small operating range and flight duration. The limited spatial scope of RECONASS and INACHUS matched their abilities well, but damage assessment over larger affected areas requires different solutions. Off-the-shelf UAVs come equipped with high quality optical cameras, though the computer vision processing to generate 3D point clouds fails for dark image patches such as shadow or smaller building openings. For this reason, openings and possible survival spaces in the pilot structures could not be mapped, and here active sensors have a clear advantage. Commercial UAVs also tend to be closed and largely proprietary systems, meaning that it is not easily possible, if at all, to exchange or add sensors, or to install processing units such as a DJI Manifold (China) or NVIDIA Jetson TX2 (USA) to push more autonomy in onboard image processing or dynamic flight path adjustment onto the drone. Several of these limitations are the focus of ongoing research, as explained in the following section.

4. Outlook and New Developments

The literature reviewed in this paper mirrors a rapidly developing discipline that in only a few years moved from largely descriptive imaging of disaster scenes to fully automated analysis procedures that build on state-of-the-art methods originating, in particular, in the computer science domain. At the same time, limits persist in hard- and software, in operational damage mapping procedures, but also in the conceptual basis of how images can be related to the actual meaning and significance of damage, which are addressed in this section.

4.1. Improvements in Machine Learning

For all the sophistication of machine leaning approaches to recognize patterns and features, some open questions persist. The black box nature of deep learning approaches means that the specific effect of certain training labels remains unclear, challenging efforts to optimize the training efficiency for specific damage features. Training to map only specific indicators such as cracks or object dislocations is thus challenging, compounded by the scarcity of large training samples for individual damage features. Also, solutions developed to date still tend to be patch/grid-based, highlighting damage in general, but not specific features. This, however, is highly scale dependent, with high resolution image data, for example, also yielding small superpixels that allow precise damage identification [50].

Work such as in [56,57] tends to focus on activation layers that indicate the presence and approximate position of damage (Figure 6), rather than the creation of actual damage maps. From a user perspective, more clarity on the specific damage type, but also more precise location, shape, and size, would be preferable. In addition, the nature of CNN-based studies prevents insights into how specifically a network with superior overall accuracy performs in terms of reducing false positives or negatives.

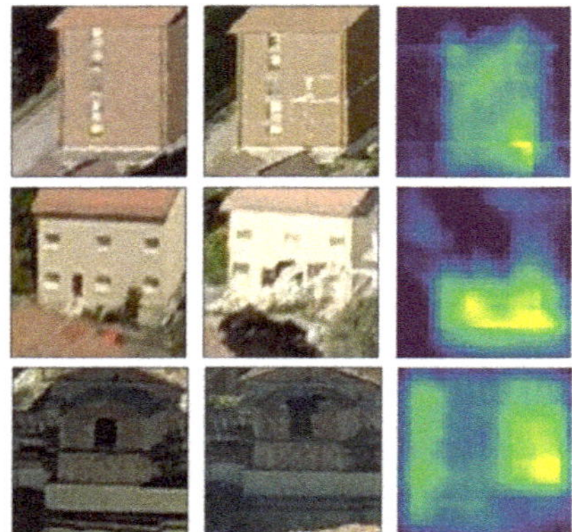

Figure 6. Examples of building damage detection via CNN activation layers, using aerial pre- and post-earthquake façade images. Bright activation colors show damage hotspots (adapted from [61]).

To overcome the problem of the large number of training samples needed in CNN analysis, recent work has shown how Generative Adversarial Networks (GAN) can effectively enlarge sample databases, which has already been shown to benefit the identification of damage of road furniture [60]. GAN seem to be particularly useful in anomaly detection [80], where training does not focus on a potentially large number of specific damage features or indicators, but rather where a comprehensive understanding of normal, undamaged scenes is created, based on which anomalies such as damage are identified. GAN have been mainly used in applications with smaller variabilities than are typical for urban scenes (i.e., indoor environments with fixed cameras). Their use in urban scenes is, therefore, an additional challenge that could be compensated by only using very large and comprehensive datasets of undamaged scenes, to prevent the generation of many false positives.

4.2. Mapping Autonomy

Traditional UAV surveys were based on pre-defined flight plans or manual piloting supported by video streams from the instrument, with data getting processed after landing of the aircraft, or through pipelines such as described in [59]. A more ideal scenario would be for the UAV to carry out an initial, for example vertical, survey over a pre-defined area, identify hotspot and damage candidate areas based on limited real-time processing, followed by a more detailed and multi-perspective survey of those marked areas. The work in [51] showed how data from an initial coarse vertical survey can be used to guide a more local assessment. Such a procedure can be implemented based on streamed data that are processed in near-real time, and adjusted flight path instructions uploaded. Alternatively, data can be processed on the UAV itself. Work described in [81,82] showed how even microdrones can perform analysis based on deep neural networks to facilitate autonomous navigation. UAVs with greater payloads have been fitted with more powerful computing units, such as NVIDIA Jetson TX2, which are capable of facilitating advanced real-time object tracking [83] or image segmentation [84].

In follow-up work to INACHUS, H2020 project PANOPTIS (Development of a Decision Support System for increasing the Resilience of Transportation Infrastructure; www.panoptis.eu) focuses on road surface and road corridor damage assessment to detect signs of gradual wear and decay, as well as the ability to respond rapidly to a disaster situation. This is done with a hybrid UAV platform (DeltaQuad from Vertical Solutions) that allows both corridor mapping of a fixed-wing platform and hovering for detailed mapping. Also, here a Jetson TX2 will be used to advance data processing on the drone itself, for both navigation and damage detection.

4.3. Indoor Mapping

UAVs have brought structural damage mapping within touching distance of buildings. Nevertheless, critical damage evidence is frequently hidden from sight, e.g., where internal load-carrying structures are compromised. In addition, damage assessment, such as defined in INACHUS, also includes support for first responders in the search for victims or survivors trapped internally, though different cavity mapping strategies only had limited success. Even with a TLS, interior cavities with connections to the outside could only be detected to a limited extent (Figure 7).

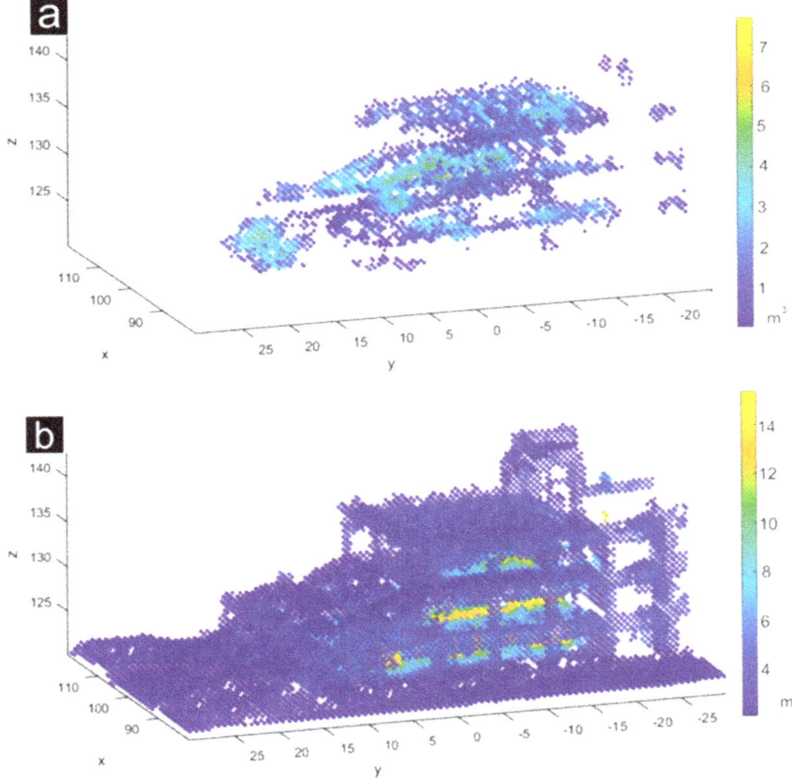

Figure 7. Voids within the photogrammetric model shown in Figure 4, obtained with a terrestrial laser scanning system. (**a**) Estimated size of open spaces observed through openings, (**b**) distance of voids to the edge of the building.

Recent work has demonstrated UAVs operating increasingly autonomously and effectively in interior, largely-GPS-denied spaces [85]. There has been a surge in research on UAV-based indoor mapping, both with single platforms and swarms. Most make use of visual SLAM to map their GPS-denied environment (e.g., [86,87]), or focusing on continuity mapping when transiting between outdoor and indoor places [88]. Others have experimented with localizing via sensors such as ultrasound [89], and the works cited in 4.2 on autonomous navigation and mapping are also relevant here. One element of improved indoor 3D reconstruction and damage mapping will be a more effective use of artificial lighting that, for example, improved the detection of small cracks in [44]. Another line of research has focused on the engineering of UAV platforms that can change shape to facilitate their entering and operating in tight spaces [90].

The damage detection work of INACHUS will also be advanced in indoor environments with H2020 project INGENIOUS (The First Responder of the Future: A Next Generation Integrated Toolkit for Collaborative Response, increasing protection and augmenting operational capacity; www.ingenious-firstresponders.eu). The focus will be on the use of drone swarms for indoor mapping to support first responders in unknown and potentially dark, smoke-filled, and hazardous indoor settings, using UAV platforms of different sizes and with different sensor load and ability, with focus on collaboration and optimization.

4.4. The Age of Drones with Robotic Abilities

UAVs tend to be fragile, susceptible to wind, and, thanks to inexpensive GPS and IMU components subject to positional inaccuracies, best operated away from structures. However, better platform control, use of collision avoidance through use of sensors or depth sensing, as well as progress in robotics and mechatronics have resulted in novel research directions. For example, in the age of aging infrastructure efforts have been spreading towards UAV-supported maintenance. This implies a number of challenges. Infrastructure is diverse and includes complicated indoor spaces such as chimneys [91], but also roads, tunnels, and bridges. Solutions are emerging to carry out day-to-day monitoring to detect defects or signs of decay, but also damage after a disaster event or accident (e.g., [92]). Such works increasingly extend into another emerging line of development, blending UAV-based abilities with robotics and mechatronics solutions. Here, UAVs are not only used to map and model infrastructure spaces, but also to carry actuator arms to place sensors for in-situ measurements [93,94], interact with objects [95,96], perform physical tests [97], or to carry out limited repairs.

5. Conclusions

Structural damage mapping with remote sensing has been a continuous research problem for decades, and for rapid operational disaster response, such as through the Charter or Copernicus EMS, reliable automated methods continue to be lacking. However, substantial progress has been made in the last decade that resulted primarily in rapid developments in UAV technology, computer vision, and in advanced image data processing with machine learning, in particular deep learning with CNN, all of which was assessed in this review. This includes a detailed analysis of the progress in image-based damage mapping that has moved from providing largely descriptive overview imagery to automated scene mapping with advanced machine learning.

The paper has shown how image-derived 3D point clouds allow a highly detailed and accurate scene reconstruction, and how the coupling of the geometric information with the original image information allows very advanced feature recognition. Classifier training is also starting to overcome the challenge of, in particular, CNN-based methods, requiring millions of training samples. The development of unsupervised CNN approaches (such as Auto-encoders) or Generative Adversarial Networks (GAN) could represent a step forward in this direction. Newer approaches are improving the efficiency, but also the transferability of classifiers, critical to be able to respond quickly to a disaster event. Comprehensive tests with first responders and urban search and rescue personnel showed that, in particular, solutions with light-weight off-the-shelf drones strike a very good compromise of high information quality and ready usability.

Developments continue at a rapid pace, with significant research efforts now being focused on UAV-based mapping in indoor settings, on UAVs also being equipped with mechatronic abilities to allow the deployment of additional sensors or to carry out repairs, though newer networks also allow more sophisticated and robust deep learning solutions. Nevertheless, more effort is needed to understand better the actual meaning and significance of specific damage evidence. In addition, UAVs need to become more autonomous to increase the efficiency of damage mapping operations. Finally, progress in the processing of UAV-based imagery, in particular through advanced machine learning, must eventually lead to fully automated and accurate damage mapping with optical satellite imagery.

Author Contributions: Conceptualization, N.K..; Methodology, N.K.; Formal Analysis, N.K., F.N., M.G., D.D., A.V.; Investigation, N.K., F.N., M.G., D.D., A.V.; Writing—Original Draft Preparation, N.K.; Writing—Review & Editing, N.K., F.N., M.G., D.D., A.V.; Visualization, N.K., F.N., M.G., D.D., A.V.; Supervision, N.K.; Funding Acquisition, N.K. All authors have read and agreed to the published version of the manuscript.

Funding: This work was funded by the EU-FP7 projects RECONASS (grant no. 312718) and INACHUS (grant no. 607522), as well as H2020 project PANOPTIS (grant no. 769129).

Acknowledgments: We thank Pictometry, Inc. for providing the Haiti and Italy imagery used in this study, and the DigitalGlobe Foundation (www.digitalglobefoundation.com) for providing satellite images on Italy and Ecuador. We also appreciate the comments made by three anonymous reviewers.

Conflicts of Interest: The authors declare no conflict of interest.

References

1. Baker, S. San francisco in ruins: The 1906 aerial photographs of george r. Lawrence. *Landscape* **1989**, *30*, 9–14.
2. Kerle, N. Disasters: Risk assessment, management, and post-disaster studies using remote sensing. In *Remote Sensing of Water Resources, Disasters, and Urban Studies (Remote Sensing Handbook, 3)*; Thenkabail, P.S., Ed.; CRC Press: Boca Raton, FL, USA, 2015; pp. 455–481.
3. Dong, L.G.; Shan, J. A comprehensive review of earthquake-induced building damage detection with remote sensing techniques. *ISPRS-J. Photogramm. Remote Sens.* **2013**, *84*, 85–99. [CrossRef]
4. Belabid, N.; Zhao, F.; Brocca, L.; Huang, Y.B.; Tan, Y.M. Near-real-time flood forecasting based on satellite precipitation products. *Remote Sens.* **2019**, *11*, 252. [CrossRef]
5. Novikov, G.; Trekin, A.; Potapov, G.; Ignatiev, V.; Burnaev, E. Satellite imagery analysis for operational damage assessment in emergency situations. In *International Conference on Business Information Systems, Berlin, Germany, 2018*; Springer International Publishing: Berlin, Germany, 2008; pp. 347–358.
6. Kerle, N.; Hoffman, R.R. Collaborative damage mapping for emergency response: The role of cognitive systems engineering. *Nat. Hazards Earth Syst. Sci.* **2013**, *13*, 97–113. [CrossRef]
7. Ghaffarian, S.; Kerle, N.; Filatova, T. Remote sensing-based proxies for urban disaster risk management and resilience: A review. *Remote Sens.* **2018**, *10*, 1760. [CrossRef]
8. Lu, C.H.; Ni, C.F.; Chang, C.P.; Yen, J.Y.; Chuang, R.Y. Coherence difference analysis of sentinel-1 sar interferogram to identify earthquake-iinduced disasters in urban areas. *Remote Sens.* **2018**, *10*, 1318. [CrossRef]
9. Li, L.L.; Liu, X.G.; Chen, Q.H.; Yang, S. Building damage assessment from polsar data using texture parameters of statistical model. *Comput. Geosci.* **2018**, *113*, 115–126. [CrossRef]
10. Gokon, H.; Post, J.; Stein, E.; Martinis, S.; Twele, A.; Muck, M.; Geiss, C.; Koshimura, S.; Matsuoka, M. A method for detecting buildings destroyed by the 2011 tohoku earthquake and tsunami using multitemporal terrasar-x data. *IEEE Geosci. Remote Sens. Lett.* **2015**, *12*, 1277–1281. [CrossRef]
11. Bai, Y.B.; Gao, C.; Singh, S.; Koch, M.; Adriano, B.; Mas, E.; Koshimura, S. A framework of rapid regional tsunami damage recognition from post-event terrasar-x imagery using deep neural networks. *IEEE Geosci. Remote Sens. Lett.* **2018**, *15*, 43–47. [CrossRef]
12. Cooner, A.J.; Shao, Y.; Campbell, J.B. Detection of urban damage using remote sensing and machine learning algorithms: Revisiting the 2010 haiti earthquake. *Remote Sens.* **2016**, *8*, 868. [CrossRef]
13. Ji, M.; Liu, L.; Buchroithner, M. Identifying collapsed buildings using post-earthquake satellite imagery and convolutional neural networks: A case study of the 2010 haiti earthquake. *Remote Sens.* **2018**, *10*, 1689. [CrossRef]
14. Xu, J.Z.; Lu, W.; Li, Z.; Khaitan, P.; Zaytseva, V. Building damage detection in satellite imagery using convolutional neural networks. *arXiv* **2019**, arXiv:1910.06444.
15. Sublime, J.; Kalinicheva, E. Automatic post-disaster damage mapping using deep-learning techniques for change detection: Case study of the tohoku tsunami. *Remote Sens.* **2019**, *11*, 1123. [CrossRef]
16. Ji, M.; Liu, L.; Du, R.; Buchroithner, M.F. A comparative study of texture and convolutional neural network features for detecting collapsed buildings after earthquakes using pre- and post-event satellite imagery. *Remote Sens.* **2019**, *11*, 1202. [CrossRef]
17. Gupta, R.; Goodman, B.; Patel, N.; Hosfelt, R.; Sajeev, S.; Heim, E.; Doshi, J.; Lucas, K.; Choset, H.; Gaston, M.E. Creating xbd: A dataset for assessing building damage from satellite imagery. In Proceedings of the CVPR Workshops, Long Beach, CA, USA, 16–20 June 2019.

18. Nakanishi, H.; Inoue, K. *Study on Intelligent Aero-Robot for Disaster Response*; Science Press Beijing: Beijing, China, 2005; Volume 5, pp. 1730–1734.
19. Miyano, K.; Shinkuma, R.; Mandayam, N.B.; Sato, T.; Oki, E. Utility based scheduling for multi-uav search systems in disaster-hit areas. *IEEE Access* **2019**, *7*, 26810–26820. [CrossRef]
20. Ejaz, W.; Azam, M.A.; Saadat, S.; Iqbal, F.; Hanan, A. Unmanned aerial vehicles enabled iot platform for disaster management. *Energies* **2019**, *12*, 2706. [CrossRef]
21. Kerle, N.; Nex, F.; Duarte, D.; Vetrivel, A. Uav-based structural damage mapping—Results from 6 years of research in two european projects. *Int. Arch. Photogramm. Remote Sens. Spatial Inf. Sci.* **2019**, *XLII-3/W8*, 187–194. [CrossRef]
22. Whang, S.H.; Kim, D.H.; Kang, M.S.; Cho, K.; Park, S.; Son, W.H. *Development of a Flying Robot System for Visual Inspection of Bridges*; Ishmii-Int Soc Structural Health Monitoring Intelligent Infrastructure: Winnipeg, MB, Canada, 2007.
23. Pollefeys, M.; Van Gool, L.; Vergauwen, M.; Verbiest, F.; Cornelis, K.; Tops, J.; Koch, R. Visual modeling with a hand-held camera. *Int. J. Comput. Vision* **2004**, *59*, 207–232. [CrossRef]
24. Suzuki, T.; Miyoshi, D.; Meguro, J.; Amano, Y.; Hashizume, T.; Sato, K.; Takiguchi, J. Real-time hazard map generation using small unmanned aerial vehicle. In Proceedings of the Sice Annual Conference, Tokyo, Japan, 20–22 August 2018; IEEE: New York, NY, USA, 2008; Volumes 1–7, p. 4.
25. Bendea, H.; Boccardo, P.; Dequal, S.; Giulio Tonolo, F.M.; Marenchino, D.; Piras, M. Low cost uav for post-disaster assessment. In *XXIst ISPRS Congress, Beijing, China*; ISPRS: Beijing, China, 2008; pp. 1373–1380.
26. Lewis, G. Evaluating the use of a low-cost unmanned aerial vehicle platform in acquiring digital imagery for emergency response. In *Geomatics Solutions for Disaster Management*; Li, J., Zlatanova, S., Fabbri, A.G., Eds.; Springer: Berlin/Heidelberg, Germany, 2007; pp. 117–133.
27. Murphy, R.R.; Steimle, E.; Griffin, C.; Cullins, C.; Hall, M.; Pratt, K. Cooperative use of unmanned sea surface and micro aerial vehicles at hurricane wilma. *J. Field Robot.* **2008**, *25*, 164–180. [CrossRef]
28. Kochersberger, K.; Kroeger, K.; Krawiec, B.; Brewer, E.; Weber, T. Post-disaster remote sensing and sampling via an autonomous helicopter. *J. Field Robot.* **2014**, *31*, 510–521. [CrossRef]
29. Adams, S.M.; Levitan, M.L.; Friedland, C.J. High resolution imagery collection for post-disaster studies utilizing unmanned aircraft systems (uas). *Photogramm. Eng. Remote Sens.* **2014**, *80*, 1161–1168. [CrossRef]
30. Dominici, D.; Baiocchi, V.; Zavino, A.; Alicandro, M.; Elaiopoulos, M. Micro uav for post seismic hazards surveying in old city center of l'aquila. In *FIG Working Week*; FIG: Rome, Italy, 2012; p. 15.
31. Mavroulis, S.; Andreadakis, E.; Spyrou, N.I.; Antoniou, V.; Skourtsos, E.; Papadimitriou, P.; Kasssaras, I.; Kaviris, G.; Tselentis, G.A.; Voulgaris, N.; et al. Uav and gis based rapid earthquake-induced building damage assessment and methodology for ems-98 isoseismal map drawing: The june 12, 2017 mw 6.3 lesvos (northeastern aegean, greece) earthquake. *Int. J. Disaster Risk Reduct.* **2019**, *37*, 20. [CrossRef]
32. Dominici, D.; Alicandro, M.; Massimi, V. Uav photogrammetry in the post-earthquake scenario: Case studies in l'aquila. *Geomat. Nat. Hazards Risk* **2017**, *8*, 87–103. [CrossRef]
33. Hein, D.; Kraft, T.; Brauchle, J.; Berger, R. Integrated uav-based real-time mapping for security applications. *ISPRS Int. Geo-Inf.* **2019**, *8*, 219. [CrossRef]
34. Xu, Z.Q.; Yang, J.S.; Peng, C.Y.; Wu, Y.; Jiang, X.D.; Li, R.; Zheng, Y.; Gao, Y.; Liu, S.; Tian, B.F. Development of an uas for post-earthquake disaster surveying and its application in ms7.0 lushan earthquake, sichuan, china. *Comput. Geosci.* **2014**, *68*, 22–30. [CrossRef]
35. Gowravaram, S.; Tian, P.Z.; Flanagan, H.; Goyer, J.; Chao, H.Y. Uas-based multispectral remote sensing and ndvi calculation for post disaster assessment. In Proceedings of the 2018 International Conference on Unmanned Aircraft Systems, Dallas, TX, USA, 12–15 June 2018; IEEE: New York, NY, USA, 2018; pp. 684–691.
36. Fernandez Galarreta, J.; Kerle, N.; Gerke, M. Uav-based urban structural damage assessment using object-based image analysis and semantic reasoning. *Nat. Hazards Earth Syst. Sci.* **2015**, *15*, 1087–1101. [CrossRef]
37. Grenzdorffer, G.J.; Guretzki, M.; Friedlander, I. Photogrammetric image acquisition and image analysis of oblique imagery. *Photogramm. Record* **2008**, *23*, 372–386. [CrossRef]
38. Gerke, M.; Kerle, N. Automatic structural seismic damage assessment with airborne oblique pictometry © imagery. *Photogramm. Eng. Remote Sens.* **2011**, *77*, 885–898. [CrossRef]

39. Gerke, M.; Kerle, N. Graph matching in 3d space for structural seismic damage assessment. In Proceedings of the 2011 IEEE International Conference on Computer Vision Workshops (ICCV Workshops), Barcelona, Spain, 6–13 November 2011.

40. Zeng, T.; Yang, W.N.; Li, X.D. Seismic damage information extent about the buildings based on low-altitude remote sensing images of mianzu quake-stricken areas. *Appl. Mech. Mater.* **2012**, *105–107*, 1889–1893. [CrossRef]

41. Vetrivel, A.; Gerke, M.; Kerle, N.; Vosselman, G. Identification of damage in buildings based on gaps in 3d point clouds from very high resolution oblique airborne images. *ISPRS-J. Photogramm. Remote Sens.* **2015**, *105*, 61–78. [CrossRef]

42. Li, S.; Tang, H.; He, S.; Shu, Y.; Mao, T.; Li, J.; Xu, Z. Unsupervised detection of earthquake-triggered roof-holes from uav images using joint color and shape features. *IEEE Geosci. Remote Sens. Lett.* **2015**, *12*, 1823–1827.

43. Vetrivel, A.; Gerke, M.; Kerle, N.; Vosselman, G. Segmentation of uav-based images incorporating 3d point cloud information. *Int. Arch. Photogramm. Remote Sens. Spat. Inf. Sci.* **2015**, *II-3/W4*, 261–268. [CrossRef]

44. Dorafshan, S.; Thomas, R.J.; Maguire, M. Fatigue crack detection using unmanned aerial systems in fracture critical inspection of steel bridges. *J. Bridge Eng.* **2018**, *23*, 15. [CrossRef]

45. Chen, S.Y.; Laefer, D.F.; Mangina, E.; Zolanvari, S.M.I.; Byrne, J. Uav bridge inspection through evaluated 3d reconstructions. *J. Bridge Eng.* **2019**, *24*, 15. [CrossRef]

46. Akbar, M.A.; Qidwai, U.; Jahanshahi, M.R. An evaluation of image-based structural health monitoring using integrated unmanned aerial vehicle platform. *Struct. Control. Health Monit.* **2019**, *26*, 20. [CrossRef]

47. Kakooei, M.; Baleghi, Y. Fusion of satellite, aircraft, and uav data for automatic disaster damage assessment. *Int. J. Remote Sens.* **2017**, *38*, 2511–2534. [CrossRef]

48. Vetrivel, A.; Duarte, D.; Nex, F.; Gerke, M.; Kerle, N.; Vosselman, G. Potential of multi-temporal oblique airborne imagery for structural damage assessment. In Proceedings of the XXIII ISPRS Congress, Commission III, International Archives of the Photogrammetry Remote Sensing and Spatial Information Sciences, Prague, Czech Republic, 12–19 July 2016; Halounova, L., Schindler, K., Limpouch, A., Pajdla, T., Safar, V., Mayer, H., Elberink, S.O., Mallet, C., Rottensteiner, F., Bredif, M., et al., Eds.; International Society for Photogrammetry and Remote Sensing: Prague, Czech Republic, 2016; Volume 3, pp. 355–362.

49. Tu, J.H.; Sui, H.G.; Feng, W.Q.; Jia, Q. Detecting facade damage on moderate damaged type from high-resolution oblique aerial images. *IEEE J. Sel. Top. Appl. Earth Obs. Remote Sens.* **2017**, *10*, 5598–5607. [CrossRef]

50. Lucks, L.; Bulatov, D.; Thönnessen, U.; Böge, M. Superpixel-wise assessment of building damage from aerial images. In Proceedings of the 14th International Joint Conference on Computer Vision, Imaging and Computer Graphics Theory and Applications, VISIGRAPP 2019, Prague, Czech Republic, 25–27 February 2019; pp. 211–220.

51. Duarte, D.; Nex, F.; Kerle, N.; Vosselman, G. Towards a more efficient detection of earthquake induced façade damages using oblique uav imagery. *Int. Arch. Photogramm. Remote Sens. Spatial Inf. Sci.* **2017**, *XLII-2/W6*, 93–100. [CrossRef]

52. Dorafshan, S.; Maguire, M. Bridge inspection: Human performance, unmanned aerial systems and automation. *J. Civ. Struct. Health Monit.* **2018**, *8*, 443–476. [CrossRef]

53. Dorafshan, S.; Coopmans, C.; Thomas, R.J.; Maguire, M. Deep learning neural networks for suas-assisted structural inspections: Feasibility and application. In Proceedings of the 2018 International Conference on Unmanned Aircraft Systems, Dallas, TX, USA, 12–15 June 2018; IEEE: New York, NY, USA, 2018; pp. 874–882.

54. Xu, Z.H.; Wu, L.X.; Zhang, Z.X. Use of active learning for earthquake damage mapping from uav photogrammetric point clouds. *Int. J. Remote Sens.* **2018**, *39*, 5568–5595. [CrossRef]

55. Vetrivel, A.; Gerke, M.; Kerle, N.; Nex, F.; Vosselman, G. Disaster damage detection through synergistic use of deep learning and 3d point cloud features derived from very high resolution oblique aerial images, and multiple-kernel-learning. *ISPRS-J. Photogramm. Remote Sens.* **2018**, *140*, 45–59. [CrossRef]

56. Duarte, D.; Nex, F.; Kerle, N.; Vosselman, G. Satellite image classification of building damages using airborne and satellite image samples in a deep learning approach. *ISPRS Ann. Photogramm. Remote Sens. Spatial Inf. Sci.* **2018**, *IV-2*, 89–96. [CrossRef]

57. Duarte, D.; Nex, F.; Kerle, N.; Vosselman, G. Multi-resolution feature fusion for image classification of building damages with convolutional neural networks. *Remote Sens.* **2018**, *10*, 1636. [CrossRef]

58. Kerle, N.; Ghaffarian, S.; Nawrotzki, R.; Leppert, G.; Lech, M. Evaluating resilience-centered development interventions with remote sensing. *Remote Sens.* **2019**, *11*, 2511. [CrossRef]

59. Nex, F.; Duarte, D.; Steenbeek, A.; Kerle, N. Towards real-time building damage mapping with low-cost uav solutions. *Remote Sens.* **2019**, *11*, 287. [CrossRef]

60. Tsai, Y.C.; Wei, C.C. *Accelerated Disaster Reconnaissance Using Automatic Traffic Sign Detection with UAV and AI*; Amer Soc Civil Engineers: New York, NY, USA, 2019; pp. 405–411.

61. Duarte, D.; Nex, F.; Kerle, N.; Vosselman, G. Detection of seismic façade damages with multi-temporal oblique aerial imagery. *GISci. Remote Sens..* In press.

62. Li, Y.D.; Ye, S.; Bartoli, I. Semisupervised classification of hurricane damage from postevent aerial imagery using deep learning. *J. Appl. Remote Sens.* **2018**, *12*, 045008. [CrossRef]

63. Li, Y.D.; Hu, W.; Dong, H.; Zhang, X.Y. Building damage detection from post-event aerial imagery using single shot multibox selector. *Appl. Sci.* **2019**, *9*, 1128. [CrossRef]

64. Liang, X. Image-based post-disaster inspection of reinforced concrete bridge systems using deep learning with bayesian optimization. *Comput.-Aided Civ. Infrastruct. Eng.* **2019**, *34*, 415–430. [CrossRef]

65. Nex, F.; Duarte, D.; Tonolo, F.G.; Kerle, N. Structural building damage detection with deep learning: Assessment of a state-of-the-art cnn in operational conditions. *Remote Sens.* **2019**, *11*, 2765. [CrossRef]

66. Song, D.M.; Tan, X.; Wang, B.; Zhang, L.; Shan, X.J.; Cui, J.Y. Integration of super-pixel segmentation and deep-learning methods for evaluating earthquake-damaged buildings using single-phase remote sensing imagery. *Int. J. Remote Sens.* **2019**, *41*, 1040–1066. [CrossRef]

67. Huang, H.; Sun, G.Y.; Zhang, X.M.; Hao, Y.L.; Zhang, A.Z.; Ren, J.C.; Ma, H.Z. Combined multiscale segmentation convolutional neural network for rapid damage mapping from postearthquake very high-resolution images. *J. Appl. Remote Sens.* **2019**, *13*, 022007. [CrossRef]

68. Vetrivel, A.; Gerke, M.; Kerle, N.; Vosselman, G. Identification of structurally damaged areas in airborne oblique images using a visual-bag-of-words approach. *Remote Sens.* **2016**, *8*, 231. [CrossRef]

69. Gong, L.X.; Wang, C.; Wu, F.; Zhang, J.F.; Zhang, H.; Li, Q. Earthquake-induced building damage detection with post-event sub-meter vhr terrasar-x staring spotlight imagery. *Remote Sens.* **2016**, *8*, 887. [CrossRef]

70. Szegedy, C.; Liu, W.; Jia, Y.Q.; Sermanet, P.; Reed, S.; Anguelov, D.; Erhan, D.; Vanhoucke, V.; Rabinovich, A. Going deeper with convolutions. In Proceedings of the 2015 IEEE Conference on Computer Vision and Pattern Recognition, Boston, MA, USA, 7–12 June 2015; IEEE: New York, NY, USA, 2015; pp. 1–9.

71. Cusicanqui, J.; Kerle, N.; Nex, F. Usability of aerial video footage for 3-d scene reconstruction and structural damage assessment. *Nat. Hazards Earth Syst. Sci.* **2018**, *18*, 1583–1598. [CrossRef]

72. Mitomi, H.; Yamzaki, F.; Matsuoka, M. Automated detection of building damage due to recent earthquakes using aerial television image. In *21st Asian Conference on Remote Sensing, Taipei, Taiwan, 2000*; GIS Development: Taipei, Taiwan, 2000; pp. 401–406.

73. Grünthal, G. *European Macroseismic Scale 1998 (ems-98)*; Cahiers du Centre Européen de Géodynamique et de Séismologie, Centre Européen de Géodynamique et de Séismologie: Walferdange, Luxembourg, 1998; Volume 15, p. 99.

74. Yamazaki, F.; Yano, Y.; Matsuoka, M. Visual damage interpretation of buildings in bam city using quickbird images following the 2003 bam, iran, earthquake. *Earthq. Spectra* **2005**, *21*, S328–S336. [CrossRef]

75. Corbane, C.; Saito, K.; Dell'Oro, L.; Gill, S.P.D.; Piard, B.E.; Huyck, C.K.; Kemper, T.; Lemoine, G.; Spence, R.J.S.; Shankar, R.; et al. A comprehensive analysis of building damage in the 12 january 2010 mw7 haiti earthquake using high resolution satellite and aerial imagery. *Photogramm. Eng. Remote Sens.* **2011**, *77*, 997–1009. [CrossRef]

76. Dubois, D.; Lepage, R. Fast and efficient evaluation of building damage from very high resolution optical satellite images. *IEEE J. Sel. Top. Appl. Earth Obs. Remote Sens.* **2014**, *7*, 4167–4176. [CrossRef]

77. Li, W.; Shuai, X.; Liu, Q. Building damage characteristics analysis based on the three-dimensional image from oblique photography. *J. Nat. Disasters* **2016**, *25*, 152–158.

78. Lattanzi, D.; Miller, G.R. 3d scene reconstruction for robotic bridge inspection. *J. Infrastruct. Syst.* **2015**, *21*, 12. [CrossRef]

79. Zou, Y.; Gonzalez, V.; Lim, J.; Amor, R.; Guo, B.; Babaeian Jelodar, M. Systematic framework for post-earthquake bridge inspection through uav and 3d bim reconstruction. In Proceedings of the CIB World Building Congress, Hong Kong, China, 17–21 June 2019; p. 9.

80. Akcay, S.; Atapour Abarghouei, A.; Breckon, T. Skip-ganomaly: Skip connected and adversarially trained encoder-decoder anomaly detection. In Proceedings of the International Joint Conference on Neural Networks (IJCNN), Budapest, Hungary, 14–19 July 2019; pp. 1–8.

81. Loquercio, A.; Kaufmann, E.; Ranftl, R.; Dosovitskiy, A.; Koltun, V.; Scaramuzza, D. Deep drone racing: From simulation to reality with domain randomization. *IEEE Trans. Robot.* **2019**, 1–14. [CrossRef]

82. Palossi, D.; Loquercio, A.; Conti, F.; Flamand, E.; Scaramuzza, D.; Benini, L. A 64mw dnn-based visual navigation engine for autonomous nano-drones. *IEEE Internet Things J.* **2019**, *6*, 8357–8371. [CrossRef]

83. Wu, H.H.; Zhou, Z.; Feng, M.; Yan, Y.; Xu, H.; Qian, L. Real-time single object detection on the uav. In Proceedings of the 2019 International Conference on Unmanned Aircraft Systems, ICUAS 2019, Atlanta, GA, USA, 11–14 June 2019; pp. 1013–1022.

84. Siam, M.; Eikerdawy, S.; Gamal, M.; Abdel-Razek, M.; Jagersand, M.; Zhang, H. Real-time segmentation with appearance, motion and geometry. In Proceedings of the IEEE International Conference on Intelligent Robots and Systems, Madrid, Spain, 1–5 October 2018; pp. 5793–5800.

85. Delmerico, J.; Mintchev, S.; Giusti, A.; Gromov, B.; Melo, K.; Horvat, T.; Cadena, C.; Hutter, M.; Ijspeert, A.; Floreano, D.; et al. The current state and future outlook of rescue robotics. *J. Field Robot.* **2019**, *36*, 1171–1191. [CrossRef]

86. Trujillo, J.C.; Munguia, R.; Guerra, E.; Grau, A. Visual-based slam configurations for cooperative multi-uav systems with a lead agent: An observability-based approach. *Sensors* **2018**, *18*, 4243. [CrossRef] [PubMed]

87. Bavle, H.; Sanchez-Lopez, J.L.; de la Puente, P.; Rodriguez-Ramos, A.; Sampedro, C.; Campoy, P. Fast and robust flight altitude estimation of multirotor uavs in dynamic unstructured environments using 3d point cloud sensors. *Aerospace* **2018**, *5*, 94. [CrossRef]

88. Zhang, X.; Xian, B.; Zhao, B.; Zhang, Y. Autonomous flight control of a nano quadrotor helicopter in a gps-denied environment using on-board vision. *IEEE Trans. Ind. Electron.* **2015**, *62*, 6392–6403. [CrossRef]

89. Paredes, J.A.; Alvarez, F.J.; Aguilera, T.; Villadangos, J.M. 3d indoor positioning of uavs with spread spectrum ultrasound and time-of-flight cameras. *Sensors* **2018**, *18*, 89. [CrossRef]

90. Falanga, D.; Kleber, K.; Mintchev, S.; Floreano, D.; Scaramuzza, D. The foldable drone: A morphing quadrotor that can squeeze and fly. *IEEE Robot. Autom. Lett.* **2018**, *4*, 209–216. [CrossRef]

91. Quenzel, J.; Nieuwenhuisen, M.; Droeschel, D.; Beul, M.; Houben, S.; Behnke, S. Autonomous mav-based indoor chimney inspection with 3d laser localization and textured surface reconstruction. *J. Intell. Robot. Syst.* **2019**, *93*, 317–335. [CrossRef]

92. Schweizer, E.A.; Stow, D.A.; Coulter, L.L. Automating near real-time, post-hazard detection of crack damage to critical infrastructure. *Photogramm. Eng. Remote Sens.* **2018**, *84*, 76–87. [CrossRef]

93. Sanchez-Cuevas, P.J.; Ramon-Soria, P.; Arrue, B.; Ollero, A.; Heredia, G. Robotic system for inspection by contact of bridge beams using uavs. *Sensors* **2019**, *19*, 305. [CrossRef] [PubMed]

94. Jimenez-Cano, A.E.; Heredia, G.; Ollero, A. Aerial manipulator with a compliant arm for bridge inspection. In Proceedings of the 2017 International Conference on Unmanned Aircraft Systems (ICUAS), Miami, FL, USA, 13–16 June 2017; pp. 1217–1222.

95. Lin, L.; Yang, Y.; Cheng, H.; Chen, X. Autonomous vision-based aerial grasping for rotorcraft unmanned aerial vehicles. *Sensors* **2019**, *19*, 3410. [CrossRef] [PubMed]

96. Ruggiero, F.; Lippiello, V.; Ollero, A. Introduction to the special issue on aerial manipulation. *IEEE Robot. Autom. Lett.* **2018**, *3*, 2734–2737. [CrossRef]

97. Salaan, C.J.; Tadakuma, K.; Okada, Y.; Ohno, K.; Tadokoro, S. Uav with two passive rotating hemispherical shells and horizontal rotor for hammering inspection of infrastructure. In Proceedings of the 2017 IEEE/SICE International Symposium on System Integration, Taipei, Taiwan, 11–14 December 2017; pp. 769–774.

MDPI
St. Alban-Anlage 66
4052 Basel
Switzerland
Tel. +41 61 683 77 34
Fax +41 61 302 89 18
www.mdpi.com

ISPRS International Journal of Geo-Information Editorial Office
E-mail: ijgi@mdpi.com
www.mdpi.com/journal/ijgi